Lecture Notes in Mathematics 1608

Editors:
A. Dold, Heidelberg
F. Takens, Groningen

Springer
Berlin
Heidelberg
New York
Barcelona
Budapest
Hong Kong
London
Milan
Paris
Tokyo

P. Biane   R. Durrett

Lectures on Probability Theory

Ecole d'Eté de Probabilités
de Saint-Flour XXIII – 1993

Editor: P. Bernard

 Springer

Authors

Philippe Biane
Université Pierre et Marie Curie
Laboratoire de Probabilités
4, Place Jussieu
F-75252 Paris, France

Richard Durrett
Cornell University
Department of Mathematics
Ithaca, NY 14853-7901, USA

Editor

Pierre Bernard
Université Blaise Pascal
Clermont-Ferrand
Laboratoire de Mathématiques Appliquées
F-63177 Aubière, France

Cataloging-in-Publication Data applied for

Die Deutsche Bibliothek - CIP-Einheitsaufnahme

Lectures on probability theory / Ecole d'Eté de Probabilités de
Saint-Flour XXIII - 1993. P. Biane ; R. Durrett. Ed.: P.
Bernard. - Berlin ; Heidelberg ; New York ; London ; Paris ; CetE
Tokyo ; Hong Kong : Springer, 1995
  (Lecture notes in mathematics ; Vol. 1608)
  ISBN 3-540-60015-9 (Berlin ...)
  ISBN 0-387-60015-9 (New York ...)
NE: Biane, Philippe; Durrett, Richard; Bernard, Pierre [Hrsg.]; Ecole
  d'Eté de Probabilités <23, 1993, Saint-Flour>; GT

Mathematics Subject Classification (1991): 60-01, 60-06, 60G60, 60H07, 60J70,
60K35, 81-06, 81S25

ISBN 3-540-60015-9 Springer-Verlag Berlin Heidelberg New York

© Springer-Verlag Berlin Heidelberg 1995
Printed in Germany
Typesetting: Camera-ready by authors

SPIN: 10130352   46/3142-543210 - Printed on acid-free paper

INTRODUCTION

This volume contains lectures given at the Saint-Flour Summer School of Probability Theory during the period 18th August - 4th September, 1993.

We thank the authors for all the hard work they accomplished. Their lectures are a work of reference in their domain.

The School brought together 62 participants, 30 of whom gave a lecture concerning their research work.

Below you will find the list of participants and their papers.

Finally, to facilitate research concerning previous schools we give here the number of the volume of "Lecture Notes" where they can be found :

Lecture Notes in Mathematics
1971 : n° 307 - 1973 : n° 390 - 1974 : n° 480 - 1975 : n° 539 -
1976 : n° 598 - 1977 : n° 678 - 1978 : n° 774 - 1979 : n° 876 -
1980 : n° 929 - 1981 : n° 976 - 1982 : n° 1097 - 1983 : n° 1117 -
1984 : n° 1180   1985 - 1986 et 1987 . n° 1362 - 1988 : n° 1427 -
1989 : n° 1464 - 1990 : n° 1527 - 1991 : n° 1541 - 1992 : n° 1581

Lecture Notes in Statistics
1986 : n° 50

TABLE OF CONTENTS

CALCUL STOCHASTIQUE

NON-COMMUTATIF

Philippe BIANE

Sommaire

3

Introduction

Le calcul stochastique non-commutatif s'est développé depuis quelques années à la suite des travaux de Hudson et Parthasarathy [40], qui ont défini des intégrales stochastiques par rapport à trois "martingales non-commutatives" a_t^+, a_t^-, et a_t^0. Ces trois processus ne sont pas composés de variables aléatoires au sens classique, mais sont des familles d'opérateurs sur l'espace L^2 de la mesure de Wiener. Le théorème spectral permet d'interpréter les combinaisons auto-adjointes de ces opérateurs comme des variables aléatoires, et le fait qu'ils ne commutent pas entre eux, leur permet d'avoir des propriétés remarquables; par exemple, le processus $(a_t^+ + a_t^-)_{t\geq 0}$ s'interprète comme un mouvement brownien ou, plus exactement, chaque opérateur $a_t^+ + a_t^-$ est l'opérateur de multiplication par la variable aléatoire B_t, où $(B_t)_{t\geq 0}$ est le mouvement brownien canonique sur l'espace de Wiener. D'autre part, pour tout $z \in \mathbb{C}$ $(a_t^0 + za_t^+ + \bar{z}a_t^- + |z|^2 t)_{t\geq 0}$ est un processus de Poisson d'intensité $|z|^2$, au sens où ces opérateurs s'interprètent comme les opérateurs de multiplication définis par un processus de Poisson sur son espace L^2, lorsqu'on a identifié l'espace L^2 de la mesure de Wiener et celui d'un processus de Poisson au moyen des décompositions en chaos. Il est clairement impossible d'obtenir de telles propriétés en utilisant des familles de variables aléatoires au sens habituel du terme. Toutes les notions évoquées ci-dessus seront expliquées en détails dans la suite du cours.

La motivation initiale du calcul stochastique stochastique non-commutatif était d'utiliser les techniques d'équations différentielles stochastiques pour résoudre des problèmes de mécanique quantique, mais il s'est avéré que la théorie ainsi développée est intéressante par elle-même et donne un point de vue nouveau sur certains aspects des probabilités classiques. C'est ainsi que la formule d'Itô y apparaît comme intimement liée aux relations de commutation d'Heisenberg qui jouent un rôle fondamental en mécanique quantique. Un sous-produit remarquable de ce calcul stochastique est la possibilité d'obtenir, en principe, tout processus de Markov comme solution d'une équation différentielle stochastique qui a la même structure formelle que celle d'une diffusion, les mouvements browniens qui dirigent l'équation étant remplacés par une version multidimensionnelle des "processus non-commutatifs" a_t^+, a_t^-, et a_t^0 (en fait pour avoir ce résultat en toute généralité il faut affronter des problèmes analytiques qui ne sont pas encore complètement résolus). On verra à la fin du cours des exemples explicites de telles constructions, dont celui des chaînes de Markov en temps continu sur un espace d'état fini, pour lequel la théorie est complète. Il apparaît que ce sont les sauts des processus de Markov qui nécessitent l'introduction de martingales non-commutatives dans l'équation différentielle stochastique.

Le but de ce cours est de présenter, à un public de probabilistes, les bases de cette théorie qui est actuellement en plein développement. L'expérience de plusieurs exposés devant des probabilistes "classiques" m'a appris que l'évocation des "variables aléatoires non-commutatives" avait tendance à plonger l'auditoire dans la perplexité. Ma première tâche va donc être de tenter de démythifier cette notion, et pour cela je vais commencer par évoquer son origine, qui est au coeur de la mécanique quantique.

C'est à von Neumann [56] que l'on doit d'avoir dégagé le formalisme mathématique de la mécanique quantique, après les travaux de nombreux physiciens dont W.Heisenberg, E.Schrödinger, et M.Born. Le postulat de base est que tout système physique peut être décrit par un espace de Hilbert complexe H et un vecteur $\psi \in H$ de norme 1 (que l'on apelle l'état du système). Souvent, le système considéré est très simple, par exemple il peut consister en une seule particule dans un espace vide, auquel cas un choix plausible de H est l'espace $L^2(\mathbf{R}^3)$, (si on ne tient pas compte du spin) et ψ s'appelle alors la fonction d'onde de la particule. Mais rien n'empêche en théorie de traiter des système complexes composés d'un grand nombre de particules; on a alors affaire à de "gros" espaces de Hilbert, et à des fonctions d'onde compliquées que l'on ne peut pas, en général, expliciter.

A toute quantité physique du système que l'on peut mesurer au moyen d'une expérience correspond un opérateur auto-adjoint A sur H. D'après le théorème spectral il existe une mesure μ_ψ sur le spectre $\sigma(A)$ telle que $\mu_\psi(f) = < f(A)\psi, \psi >$ pour toute fonction borélienne bornée sur $\sigma(A)$, $f(A)$ étant défini par le cacul fonctionnel des opérateurs auto-adjoints. Lorsque A a un spectre discret formé de valeurs propres λ_i de multiplicité 1, cette mesure de probabilité est simple à décrire, on a $\mu_\psi(\{\lambda_i\}) = |< \psi, \phi_i >|^2$, ϕ_i étant un vecteur propre de norme 1 associé à λ_i. On postule que le résultat d'une mesure de la quantité physique correspondant à A est une variable aléatoire de loi μ_ψ lorsque le système est dans l'état ψ. En particulier, cette loi n'est une mesure de Dirac que lorsque le vecteur ψ est un vecteur propre de A. Un autre postulat ("réduction du paquet d'ondes") énonce qu'après la mesure de la quantité correspondant à l'opérateur A, si cette mesure a donné comme résultat λ (où λ est une valeur propre de A de vecteur propre correspondant ϕ), le système se trouve dans l'état ϕ.

La mécanique quantique est une théorie intrinsèquement probabiliste, car il existe des systèmes physiques pour lesquels la connaissance exacte de toutes les données (i.e. de la fonction d'onde du système) ne permet pas de prédire avec certitude le résultat d'une expérience.

Pour les lecteurs qui ne sont pas familiers avec la mécanique quantique, je vais illustrer les postulats ci-dessus par la description d'expériences de physique (comme dans Feynman [35]) inspirées de celle de Stern et Gerlach, et en donner une interprétation avec les postulats de la mécanique quantique. Cette description nous permettra de rencontrer un premier exemple de variable aléatoire "non-commutative", exemple sur lequel on reviendra au chapitre 1.

On suppose donné un dispositif produisant un faisceau linéaire horizontal de particules de même nature (par exemple des atomes d'hydrogène, ou des molécules d'azote, ...), que l'on envoie à travers un appareil qui crée un champ magnétique ayant un fort gradient vertical. A la sortie de l'appareil, on constate que le faisceau de particules s'est scindé en plusieurs sous-faisceaux présentant des déviations verticales par rapport à la direction initiale (voir *fig.* 1).

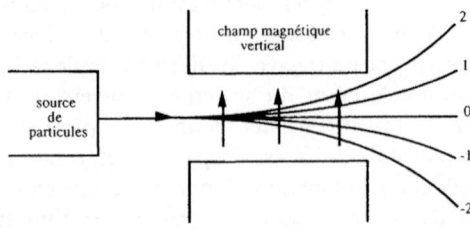

fig. 1

Les déviations de ces faisceaux par rapport à la direction initiale sont des multiples entiers d'une même quantité. Cette quantité ainsi que le nombre de faisceaux sont des caractéristiques du type des particules utilisées, elles ne dépendent pas de la direction du champ magnétique, si celui-ci reste orthogonal au faisceau. Si le faisceau initial se scinde en N faisceaux, la particule est dite de spin J où J est le demi-entier $J = \frac{N-1}{2}$ (dans le cas de la figure 1, le spin est 2). On interprète cette expérience en postulant que chaque particule possède un "moment magnétique" dont la composante suivant l'axe vertical est un multiple demi-entier d'une certaine quantité, ce demi-entier étant de la forme $-J, -J+1, \ldots, J$. (C'est cette propriété de quantification de certaines quantités physiques associées aux particules élémentaires qui a donné son nom à la mécanique quantique).

On peut modifier cette expérience en faisant passer les faisceaux à travers un second appareil, placé juste à la sortie du premier, et créant un champ magnétique opposé, de façon à remettre les particules dans leur direction initiale. En disposant judicieusement des caches entre les deux champs magnétiques, on peut filtrer les faisceaux de manière à n'en garder qu'un.

fig. 2

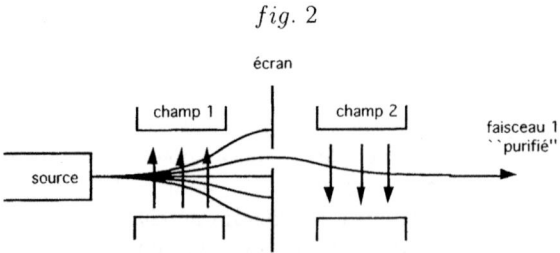

En faisant passer un tel faisceau "purifié" à travers un troisième champ vertical, on constate qu'il ne se sépare plus et qu'il dévie de la même valeur qu'auparavant. Ce procédé permet donc de sélectionner les particules suivant la valeur de leur moment magnétique vertical (*fig.* 3).

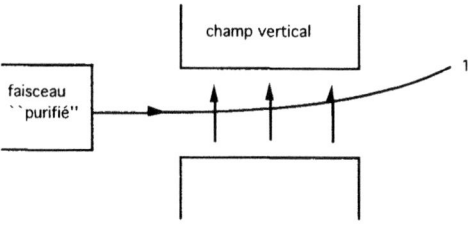

fig. 3

Afin de simplifier la discussion, je vais maintenant supposer que les particules ont un spin $J = \frac{1}{2}$, c'est à dire que le moment magnétique d'une telle particule suivant une direction donnée ne peut prendre que les valeurs $+1$ ou -1 dans une unité de mesure convenable. Au moyen du procédé décrit ci-dessus on se donne un faisceau de telles particules de moment magnétique vertical $+1$, que l'on fait passer à travers un champ magnétique faisant un angle Δ avec la verticale (*fig.* 4, $\Delta \in [0, \frac{\pi}{2}]$).

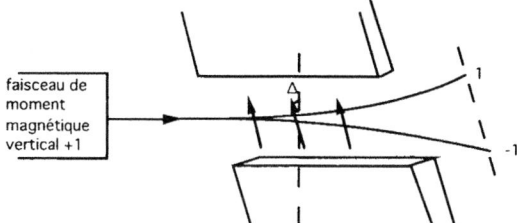

fig. 4

On constate que le faisceau s'est séparé en deux. Si l'on mesure l'intensité de chaque faisceau (c'est à dire la proportion de particules du faisceau initial dans chacun des deux faisceaux de sortie), on obtient respectivement les valeurs $\cos^2 \frac{\Delta}{2}$ et $\sin^2 \frac{\Delta}{2}$.

Utilisons un dispositif analogue à celui de la figure 2 pour sélectionner les particules de la figure 4 qui ont un moment magnétique $+1$ dans la direction Δ. Si maintenant on fait passer à nouveau ces particules dans un champ magnétique vertical, on constate que le faisceau se sépare en 2, autrement dit, après être passées dans le champ magnétique non vertical les particules ont "oublié" leur moment magnétique vertical. On peut recommencer l'expérience avec des directions arbitraires, à chaque fois le résultat est le suivant: si on a sélectionné un faisceau de particules ayant un moment magnétique $+1$ dans la direction D, et que l'on mesure leurs moments magnétiques dans la direction D' faisant un angle Δ avec D, on trouve $+1$ et -1 avec des proportions respectives de $\cos^2 \frac{\Delta}{2}$ et $\sin^2 \frac{\Delta}{2}$.

Pour essayer d'expliquer ces expériences de manière classique il faudrait pour chaque particule introduire une famille de variables aléatoires indexées par les directions de l'espace, et prenant les valeurs $\{-1, +1\}$, qui changeraient de façon compliquée par un passage au travers d'un champ magnétique. L'interprétation quantique, elle, ne fait intervenir que des vecteurs de dimension 2 et des matrices 2×2. Voici cette interprétation. L'espace de Hilbert qui décrit le moment magnétique d'une particule est de dimension 2. Chaque particule du faisceau a un état qui est un vecteur colonne $\psi = \begin{pmatrix} u \\ v \end{pmatrix}$ avec $u \in \mathbb{C}$, $v \in \mathbb{C}$. A chaque direction de l'espace $\mathbf{R}^3$, on associe une matrice 2×2, la direction de vecteur unitaire (x, y, z) étant associée à la matrice $\begin{pmatrix} z & y + ix \\ y - ix & -z \end{pmatrix}$. La mesure d'un moment magnétique dans la direction (x, y, z) correspond à l'opérateur auto-adjoint de matrice $\begin{pmatrix} z & y + ix \\ y - ix & -z \end{pmatrix}$. D'après les postulats de la mécanique quantique, si une particule est dans l'état $\psi = \begin{pmatrix} u \\ v \end{pmatrix}$, la mesure du moment magnétique dans la direction (x, y, z) est une variable aléatoire qui vaut 1 avec probabilité $| < \psi, \alpha_+ > |^2$ et -1 avec probabilité $| < \psi, \alpha_- > |^2$, où α_+ et α_- sont des vecteurs propres unitaires de $\begin{pmatrix} z & y + ix \\ y - ix & -z \end{pmatrix}$ de valeurs propres respectives $+1$ et -1. D'autre part, si le résultat de la mesure a donné $+1$ (ou -1), après cette mesure, la particule se trouve dans l'état α_+ (ou α_-).

Reprenons l'exemple des expériences décrites ci-dessus.

La mesure du moment magnétique vertical est donnée par l'opérateur $\begin{pmatrix} 1 & 0 \\ 0 & -1 \end{pmatrix}$. Les particules de moment magnétique $+1$ dans cette direction sont donc dans l'état $e_1 = \begin{pmatrix} 1 \\ 0 \end{pmatrix}$. Le moment magnétique dans la direction $(0, \sin\Delta, \cos\Delta)$ est mesuré par l'opérateur $\begin{pmatrix} \cos\Delta & \sin\Delta \\ \sin\Delta & -\cos\Delta \end{pmatrix}$, de vecteurs propres $\alpha_+ = \begin{pmatrix} \cos\frac{\Delta}{2} \\ \sin\frac{\Delta}{2} \end{pmatrix}$ $\alpha_- = \begin{pmatrix} \sin\frac{\Delta}{2} \\ \cos\frac{\Delta}{2} \end{pmatrix}$.

Le résultat de la mesure est donc une variable de Bernoulli de probabilités: $P(1) = | < e_1, \alpha_+ > |^2 = \cos^2\frac{\Delta}{2}$ et $P(-1) = | < e_1, \alpha_- > |^2 = \sin^2\frac{\Delta}{2}$, comme le montre l'expérience. Un calcul semblable avec des directions arbitraires dans l'espace permet de retrouver tous les résultats de l'expérience.

Cet exemple nous a permis de rencontrer pour la première fois des "variables de Bernoulli quantiques". Nous les étudierons plus en détails au chapitre 1 du cours.

Après cette brève incursion dans la physique des particules, nous allons nous concentrer sur l'aspect purement mathématique des probabilités quantiques. Le plan du cours suit un chemin parallèle à celui d'un cours de probabilités classique, partant des probabilités élémentaires sur un ensemble fini, passant aux variables "continues" avec le théorème central limite, puis aux processus stochastiques, (mouvement brownien, processus de Poisson), et enfin à l'intégration stochastique.

On commence au chapitre 1 par définir la notion d'espace de probabilité non-commutatif fini, et les variables aléatoires sur un tel espace. Il s'agit en fait simplement d'algèbre linéaire de dimension finie, mais l'interprétation probabiliste que l'on en fait amène à des considérations intéressantes. Alors que la théorie élémentaire des probabilités utilise essentiellement la combinatoire et le dénombrement (voir Feller [34]), ces méthodes sont remplacées ici par l'algèbre linéaire. Même dans une situation aussi simple on voit rapidement apparaître des effets "quantiques" non-triviaux comme on le montre en étudiant la marche de Bernoulli quantique, un analogue non-commutatif du jeu de pile ou face, et le "processus de spin" associé. A la fin du chapitre 1, on démontre un théorème limite qui contient les deux théorèmes limite en loi sur les sommes de variables de Bernoulli, d'une part le théorème de Moivre-Laplace, d'autre part le théorème de convergence vers la loi de Poisson. On verra que le formalisme non-commutatif permet de réunir ces deux résultats en un seul. Les variables limites obtenues joueront un rôle fondamental par la suite.

L'énoncé du théorème limite montre la nécessité de sortir du cadre de la dimension finie, pour traiter aussi bien des variables continues, comme les gaussiennes, que des variables discrètes prenant une infinité de valeurs, comme les variables de Poisson. Cela est rendu possible grâce à la théorie spectrale des opérateurs auto-adjoints dont les principaux résultats sont rappelés au chapitre 2.

Avec le chapitre 3 on rentre dans le vif du sujet en étudiant de façon approfondie les objets introduits à la fin du chapitre 1, qui sont connus en physique quantique sous le nom d'opérateurs de création, d'annihilation, et de nombre. Le chapitre 4 est consacré aux espaces de Fock qui fournissent un moyen canonique de construire des familles d'opérateurs de création, annihilation et nombre. Ce chapitre contient des notions qui sont bien connues des spécialistes du calcul de Malliavin (cf par exemple Watanabe [87]), seul le langage dans lequel elles sont exprimées est un peu différent. C'est dans ce chapitre que l'on introduit les processus a_t^+, a_t^-, et a_t^0 de Hudson et Parthasarathy.

Dans le chapitre 5, on aborde le calcul stochastique non-commutatif proprement dit, inspiré du calcul stochastique d'Itô. On verra que ce nouveau calcul stochastique est très lié à l'intégrale de Skorokhod (voir Skorokhod [80]), une extension de l'intégrale d'Itô, qui est en fait un objet purement Hilbertien, défini comme une divergence, comme l'on remarqué Gaveau et Trauber [36]. Dans le chapitre 6 on montre comment ce calcul permet de construire explicitement des processus de Markov non-commutatifs, cette notion étant une extension de la notion usuelle de processus de Markov. Dans le cas particulier des processus de Markov au sens habituel, cette construction permet d'étendre à des processus de Markov non-nécessairement continus la construction des diffusions à l'aide d'équation différentielles stochastiques. On décrira comment fonctionne ce formalisme dans le cas simple des processus de naissance et des chaînes de Markov sur un espace d'états fini.

Un dernier chapitre est consacré à des commentaires et des compléments, ainsi qu'aux références bibliographiques.

Les connaissances requises pour lire le cours se limitent à une bonne familiarité avec les notions de base de probabilité et de théorie des opérateurs (théorie spectrale), mais en pratique il serait utile d'avoir une bonne connaissance du calcul stochastique usuel comme exposé par exemple dans [17], [21] ou [69]. En particulier, les sujets suivants, populaires parmi les probabilistes, utilisent des notions très proches: calcul de Malliavin et analyse sur l'espace de Wiener, intégrale de Skorokhod, équations différentielles stochastiques. La théorie des représentations des groupes est présente implicitement dans une grande partie du cours, mais on ne l'utilise pas vraiment (sauf au paragraphe 6.2). Le point de vue des groupes est beaucoup plus présent dans le livre de Parthasarathy [60].

Je remercie vivement les organisateurs de l'Ecole d'été de Saint-Flour de m'avoir donné la possibilité de faire ce cours. Je suis également reconnaissant à tous ceux qui ont bien voulu faire des commentaires sur la première version du cours, et je remercie tout particulièrement Monique Pontier dont les suggestions ainsi que les corrections qu'elle m'a indiquées m'ont considérablement aidé, et Jacques Azéma pour sa lecture critique et méticuleuse de la première version du cours.

Chapitre 1

Espaces de probabilité non-commutatifs finis

1.1 Espaces de probabilités

Ce chapitre est consacré à la notion d'espace de probabilité non-commutatif, en commençant par l'exemple le plus simple, qui est l'analogue d'un espace de probabilité usuel fini. Comme indiqué dans l'introduction l'idée de base, qui a son origine dans la mécanique quantique, est qu'un opérateur auto-adjoint sur un espace de Hilbert peut s'interpréter comme une variable aléatoire. Pour préciser cette correspondance, commençons par montrer comment associer à une variable aléatoire un opérateur auto-adjoint.

Soit $(\Omega, \mathcal{F}, P)$ un espace de probabilité fini, avec la tribu formée de toutes les parties de Ω. L'espace des variables aléatoires complexes $L^2_{\mathbf{C}}(\Omega, \mathcal{F}, P)$ est muni de sa structure d'espace de Hilbert complexe de dimension finie. Soit X une variable aléatoire réelle sur $(\Omega, \mathcal{F}, P)$, on définit un opérateur auto-adjoint

$$T_X : L^2_{\mathbf{C}}(\Omega, \mathcal{F}, P) \to L^2_{\mathbf{C}}(\Omega, \mathcal{F}, P)$$

$$f \to Xf$$

$L^2_{\mathbf{C}}(\Omega, \mathcal{F}, P)$ a une base orthonormale $(\delta_\omega = \frac{\chi_\omega}{\sqrt{P(\{\omega\})}})_{\omega \in \Omega}$ où χ_ω désigne la fonction indicatrice de ω (on suppose que $P(\{\omega\}) > 0$ pour tout $\omega \in \Omega$). Dans cette base, l'opérateur T_X est diagonal, ses valeurs propres sont les valeurs prises par la variable aléatoire X, puisque $T_X(\delta_\omega) = X(\omega)\delta_\omega$. De plus, l'espérance de la variable aléatoire X est donnée par la formule $E[X] = <X1, 1>$ où 1 désigne la variable identiquement égale à 1.

Réciproquement, donnons nous un opérateur auto-adjoint A sur un espace de Hilbert complexe E de dimension finie , et $(\varphi_i)_{i \in I}$, $I = \{1, \ldots, n\}$ une base orthonormale de vecteurs propres de A, correspondant aux valeurs propres $\{\lambda_i\}_{i \in I}$ (avec éventuellement des répétitions en cas de valeurs propres de multiplicité > 1). Soit $u \in E$ un vecteur unitaire, il détermine une loi de probabilité μ sur $(I, \mathcal{I})$ (I est muni de la tribu maximale $\mathcal{I}$) en posant $\mu(\{i\}) = |<u, \varphi_i>|^2$, ainsi qu'une isométrie $\iota : l^2(I, \mathcal{I}, \mu) \to E$ qui envoie χ_i sur $<u, \varphi_i> \varphi_i$. Soit X la variable aléatoire sur $(I, \mathcal{I})$ définie par $X(i) = \lambda_i$, alors $A = \iota \circ T_X \circ \iota^*$, donc l'opérateur A s'interprète comme un opérateur de multiplication par la variable aléatoire X. De plus, l'espérance de X s'exprime à l'aide de A par la formule $E[X] = <Au, u>$.

Les deux exemples ci-dessus montrent comment on peut passer de la donnée de (E, u, A), E étant un espace de Hilbert u un vecteur unitaire de E et A un opérateur auto-adjoint sur E à celle de $(\Omega, \mathcal{F}, P, X)$ où $(\Omega, \mathcal{F}, P)$ est un espace de probabilités et X une variable aléatoire. Le principe des probabilités quantiques consiste à considérer la première donnée comme fondamentale, et à remplacer l'espace de probabilité usuel par un couple (E, u) et les variables aléatoires par les opérateurs auto-adjoints sur E.

Prendre l'espérance d'une variable aléatoire correspond alors à appliquer la forme linéaire $A \mapsto\ < Au, u >$. Comme en probabilités classiques, où l'on a parfois besoin de considérer plusieurs probabilités sur le même espace, on veut pouvoir considérer simultanément des vecteurs unitaires $u_1, \ldots, u_n$. Lorsqu'on fait une combinaison convexe des fonctionnelles $A \mapsto\ < Au_k, u_k >$ définies sur $\mathcal{L}(E)$ l'espace des opérateurs sur E, on n'obtient pas une fonctionnelle du type $A \mapsto\ < Au, u >$. Observons que $< Au, u >= tr(A\pi_u)$, tr étant la forme linéaire donnant la trace d'un opérateur, et π_u le projecteur orthogonal sur la droite engendrée par u. On voit donc que si $\sum_k p_k = 1$, $\sum_k p_k < Au_k, u_k >= tr(AS)$ où $S = \sum_k p_k \pi_{u_k}$ est un opérateur positif de trace 1. En fait, tout opérateur positif de trace 1 s'obtient de cette façon.

Donnons une première définition.

DÉFINITION 1 – Un espace de probabilité non-commutatif fini est la donnée d'un espace de Hilbert de dimension finie, E et d'un opérateur positif S sur E, de trace 1. La forme linéaire sur $\mathcal{L}(E)$ (l'espace des opérateurs sur E) donnée par

$$A \mapsto tr(AS)$$

est appelée l'état correspondant à S.

Par abus de langage, on confondra parfois l'opérateur S avec l'état qu'il définit. L'ensemble des états est une partie convexe compacte du dual de $\mathcal{L}(E)$, dont les points extrémaux sont les états de la forme $A \mapsto\ < Au, u >$ (appelés aussi états purs), où u est un vecteur de norme 1. Un état pur est associé à un opérateur π_u qui est le projecteur orthogonal sur la droite engendrée par u; en particulier, deux vecteurs unitaires déterminent le même état si et seulement s'ils sont colinéaires. Remarquons tout de suite que tout état est combinaison convexe d'un nombre fini d'états purs mais qu'il n'y pas unicité de la décomposition (par exemple l'état associé à $\frac{1}{dim(E)} Id$ est égal à $\frac{1}{dim(E)} \sum_i \pi_{u_i}$ pour toute base orthonormale (u_i) de E).

1.2 Variables aléatoires, lois

Après avoir introduit les espaces de probabilité, passons aux variables aléatoires.

DÉFINITION 2 – Une variable aléatoire non-commutative sur l'espace de probabilité non-commutatif fini (E, S) est un opérateur auto-adjoint sur E.

Lorsqu'aucune confusion n'est possible, on parlera de variable aléatoire tout court.

L'exemple du paragraphe 1.1 montre que toute variable aléatoire réelle au sens usuel peut être considérée comme une variable aléatoire non-commutative, en la faisant agir par multiplication sur un espace L^2. Soit A une variable aléatoire non-commutative, appelons $\sigma(A)$ son spectre, et soit f une fonction réelle sur $\sigma(A)$. On définit un opérateur autoadjoint en posant $f(A)x = f(\lambda)x$ pour tout $\lambda \in \sigma(A)$ et x vecteur propre de A de valeur propre λ.

DÉFINITION 3 – La loi de la variable A dans l'état S est la mesure de probabilité sur $\sigma(A)$ donnée par la formule:

$$\mu(f) = tr(f(A)S)$$

pour toute fonction f sur $\sigma(A)$.

Soit $\lambda \in \sigma(A)$ on a $\mu(\{\lambda\}) = tr(S\pi_\lambda)$ où π_λ est le projecteur orthogonal sur le sous-espace propre associé à λ. On vérifie immédiatement en reprenant l'exemple du paragraphe 1.1 que $T_{f(X)} = f(T_X)$ et que la loi de T_X dans l'état pur 1 (ici 1 désigne la fonction constante égale à 1) est bien la loi de la variable aléatoire X. Les notions de variable aléatoire et de loi que nous venons de définir sont donc des extensions des notions usuelles. La première différence avec la théorie des probabilités classiques apparait quand on essaie d'appliquer les opérations algébriques élémentaires. En effet, alors qu'on peut toujours faire le produit de deux variables aléatoires usuelles, le produit de deux opérateurs auto-adjoints n'est auto-adjoint que s'ils commutent. Nous allons examiner plus en détail ce dernier cas. Soient A et B deux opérateurs auto-adjoints qui commutent, on peut décomposer E en somme directe de sous-espaces propres communs à A et B. Soit $(E_{\lambda,\nu})_{\lambda \in \sigma(A), \nu \in \sigma(B)}$ une telle décomposition on peut définir $f(A, B)$ pour une fonction f sur $\sigma(A) \times \sigma(B)$ par $f(A, B)x = f(\lambda, \nu)x$ si $x \in E_{\lambda,\nu}$. Ce calcul fonctionnel respecte l'addition et la multiplication des fonctions, ce qui fait qu'on peut définir la loi jointe de A et B qui est la mesure sur $\sigma(A) \times \sigma(B)$ déterminée par

$$\mu_{A,B}(f) = tr(f(A, B)S)$$

Plus généralement, une famille $(X_t)_{t \in T}$ d'opérateurs auto-adjoints qui commutent admet un calcul fonctionnel qui permet de définir leur loi jointe.

DÉFINITION 4 – *Une famille d'opérateurs sur un espace de Hilbert, indexée par un ensemble T (le temps) est appelée un processus stochastique non-commutatif. Si ces opérateurs sont auto-adjoints et commutent, le processus est dit classique.*

DÉFINITION 5 – *La loi d'un processus classique $(X_t)_{t \in T}$ est la loi jointe des variables $(X_t)_{t \in T}$, c'est à dire la mesure de probabilité sur $\prod_{t \in T} \sigma(X_t)$ déterminée par la forme linéaire*

$$f \mapsto tr(S f((X_t)_{t \in \mathbb{R}}))$$

sur l'ensemble des fonctions cylindriques bornées sur $\prod_{t \in T} \sigma(X_t)$.

Lorsqu'on considère des opérateurs non auto-adjoints, ou qui ne commutent pas entre eux, la notion de loi jointe de ces opérateurs n'a pas de sens.

Dans la suite du cours, nous allons rencontrer de nombreuses familles d'opérateurs auto-adjoints qui commutent, et calculer les lois des processus classiques correspondants. Un des aspects remarquables de ces processus sera leur construction à partir d'opérateurs qui ne commutent pas entre eux, et pour lesquels la notion de "loi jointe" n'a pas de sens.

1.3 Espace de Bernoulli non-commutatif

Nous allons maintenant étudier plus en détail l'exemple le plus simple d'espace de probabilité non-commutatif, en considérant l'espace de Hilbert $E = \mathbb{C}^2$ avec sa base canonique (e_1, e_2), et le produit hermitien usuel.

Les opérateurs linéaires sur E sont représentés dans la base canonique par des matrices 2×2. Un état se représente par une matrice de la forme:

$$S = \begin{pmatrix} \alpha & \bar{z} \\ z & 1 - \alpha \end{pmatrix} \quad \text{avec } \alpha \in \mathbb{R} \ z \in \mathbb{C} \ \alpha(1 - \alpha) \geq z\bar{z}, \ 0 \leq \alpha \leq 1 \tag{1}$$

Lorsque $\alpha(1 - \alpha) = z\bar{z}$, S est un projecteur orthogonal sur une droite complexe, l'état associé est donc un état pur.

Les variables aléatoires forment un espace vectoriel réel de dimension 4, noté V. La formule $< A, B > = \frac{1}{2} tr(AB^*)$ définit un produit hermitien sur $M_2(\mathbb{C})$ dont la restriction à V est un produit euclidien. Une base orthogonale de V est donnée par les matrices

$$I = \begin{pmatrix} 1 & 0 \\ 0 & 1 \end{pmatrix}, \ \sigma_x = \begin{pmatrix} 0 & 1 \\ 1 & 0 \end{pmatrix}, \ \sigma_y = \begin{pmatrix} 0 & -i \\ i & 0 \end{pmatrix}, \ \sigma_z = \begin{pmatrix} 1 & 0 \\ 0 & -1 \end{pmatrix} \quad (2)$$

Les matrices σ_x, σ_y, σ_z sont connues en physique sous le nom de matrices de Pauli, et nous les avons rencontrées dans l'introduction, elles servent à décrire les particules de spin $\frac{1}{2}$. Ces matrices ne commutent pas entre elles mais vérifient les relations suivantes

$$[\sigma_x, \sigma_y] = 2i\sigma_z \quad [\sigma_z, \sigma_x] = 2i\sigma_y \quad [\sigma_y, \sigma_z] = 2i\sigma_x \quad (3)$$

(Dans la suite, la notation $[X, Y]$ désigne le commutateur $XY - YX$).

Nous aurons également besoin des matrices adjointes l'une de l'autre

$$\delta^- = \begin{pmatrix} 0 & 1 \\ 0 & 0 \end{pmatrix} \ \delta^+ = \begin{pmatrix} 0 & 0 \\ 1 & 0 \end{pmatrix} \quad (4)$$

grâce auxquelles ont peut représenter les autres

$$I = \delta^+\delta^- + \delta^-\delta^+, \ \sigma_x = \delta^+ + \delta^-, \ \sigma_y = i(\delta^+ - \delta^-) \ \sigma_z = \delta^-\delta^+ - \delta^+\delta^- \quad (5)$$

Le groupe des matrices 2×2 unitaires agit par conjugaison sur l'espace V. Pour toute matrice U unitaire, l'application $A \mapsto UAU^*$ est $\mathbb{R}$-linéaire sur V et préserve le produit des matrices, le passage à l'adjoint, et donc la forme hermitienne scalaire et les relations de commutation. Quitte à conjuguer par une telle matrice, on pourra supposer que l'état S est donné par la matrice

$$\begin{pmatrix} p & 0 \\ 0 & q \end{pmatrix} \text{ avec } p + q = 1, \ p \geq q \geq 0$$

ce que nous ferons par la suite.

Les deux cas extrémaux $p = 1$ et $p = \frac{1}{2}$ sont particulièrement intéressants. Le premier correspond à l'état pur associé au vecteur e_0, le second est une trace, c'est à dire que l'on a $tr(SAB) = tr(SBA)$ pour tout couple d'opérateurs (A, B) (c'est d'ailleurs le seul état qui vérifie cette propriété).

Terminons ce chapitre en calculant la loi d'une variable aléatoire

$$A = x\sigma_x + y\sigma_y + z\sigma_z + t\,Id \quad x, y, z, t \in \mathbb{R}$$

dans l'état S. Le spectre de A est composé de deux points,

$$\sigma(A) = \{t \pm \sqrt{x^2 + y^2 + z^2}\}$$

et la loi de A est la mesure μ telle que

$$\mu(t + \sqrt{x^2 + y^2 + z^2}) = \frac{1}{2}(1 + \frac{z}{\sqrt{x^2 + y^2 + z^2}}(p - q))$$

$$\mu(t - \sqrt{x^2 + y^2 + z^2}) = \frac{1}{2}(1 - \frac{z}{\sqrt{x^2 + y^2 + z^2}}(p - q))$$

On remarque en particulier que lorsque $p = 1$, en faisant varier (x, y, z, t) on obtient toutes les lois sur $\mathbb{R}$ portées par (au plus) deux points. Lorsque $p = \frac{1}{2} = q$, les lois obtenues sont toutes les lois uniformes portées par deux points.

L'espace (E, S) est l'analogue pour les probabilités quantiques de l'espace de probabilité d'une variable de Bernoulli en probabilités classiques. Dans tous les cours de probabilités élémentaires, une des premières choses que l'on fait avec des variables de Bernoulli c'est d'en additioner des copies indépendantes. Nous allons faire la même chose dans le paragraphe suivant, en introduisant les marches de Bernoulli quantiques.

1.4 Addition de variables de Bernoulli quantiques indépendantes

Nous allons étudier le premier exemple de processus stochastique non-commutatif qui est un analogue quantique du jeu de pile ou face. Cela nous permettra de nous familiariser avec les notions introduites jusqu'ici en faisant des calculs explicites sur un exemple non-trivial.

En probabilités classiques, il est facile de construire des copies indépendantes $X_1, \ldots, X_N$ d'une même variable aléatoire X sur $(\Omega, \mathcal{F}, P)$, en considérant l'espace produit $(\Omega^N, \mathcal{F}^{\otimes N}, P^{\otimes N})$ et en posant $X_j(\omega_1, \ldots, \omega_N) = X(\omega_j)$.

L'espaces L^2 obtenu est un produit tensoriel $L^2(\Omega^N, \mathcal{F}^{\otimes N}, P^{\otimes N}) = (L^2(\Omega, \mathcal{F}, P))^{\otimes N}$, et l'opérateur de multiplication par la variable X_j se représente sous la forme

$$T_{X_j} = I \otimes \ldots \otimes T_X \otimes \ldots \otimes I$$

l'opérateur T_X agissant sur le j^e facteur du produit tensoriel.

On va faire la même chose dans le cas non-commutatif, c'est à dire considérer l'espace de Hilbert $E^{\otimes N}$, et l'état $S^{\otimes N}$. Afin de rester avec des espaces de dimension finie, on se restreint à N fini. Si A est une variable aléatoire sur E, les opérateurs

$$A^{(k)} = I \otimes \ldots \otimes I \otimes A \otimes I \ldots I \otimes I$$
$$\uparrow$$
$$k^e \text{ place}$$

sont des variables aléatoires non-commutatives.

Si $l \neq k$, et A, B sont des variables, alors $A^{(k)}$ et $B^{(l)}$ commutent. De plus dans l'état $S^{\otimes N}$ leur loi jointe est celle de deux variables indépendantes. En effet si f et g sont deux fonctions sur $\sigma(A)$ et $\sigma(B)$ on a

$$tr\big(S^{\otimes N} f(A^{(k)}) g(B^{(l)})\big) = tr\big(S^{\otimes N}(I \otimes \ldots \otimes f(A) \otimes \ldots \otimes g(B) \otimes \ldots \otimes I)\big)$$
$$= tr(S f(A)) \, tr(S g(B))$$

d'où l'on déduit que la loi jointe de $A^{(k)}$ et $B^{(l)}$ est une loi produit.

En particulier, pour toute variable A, la famille $(A^{(k)})_{1 \leq k \leq N}$ est un processus classique dont la loi est celle d'une famille de variables indépendantes équidistribuées avec comme loi commune la loi de A.

Reprenons les notations du paragraphe 1.3 et posons

$$x_k = \sigma_x^{(k)}, y_k = \sigma_y^{(k)}, z_k = \sigma_z^{(k)}$$

$$X_n = \sum_{k=1}^{n} x_k \quad Y_n = \sum_{k=1}^{n} y_k \quad Z_n = \sum_{k=1}^{n} z_k \tag{6}$$

Les variables $(x_k)_{1 \leq k \leq N}$ (resp. y_k, z_k) commutent, sont indépendantes et suivent des lois de Bernoulli portées par $\{\pm 1\}$.

Les trois processus $(X_n)_{n \leq N}, (Y_n)_{n \leq N}, (Z_n)_{n \leq N}$, sont donc des processus classiques, dont les lois sont celles de marches de Bernoulli. En revanche, ces trois processus ne commutent pas entre eux, en fait d'après les relations de commutation (3) on a

$$[X_k, Y_l] = 2i Z_{k \wedge l} \tag{7}$$

ainsi que les relations qui s'en déduisent par permutation circulaire de X, Y, Z.

DÉFINITION 6 – On appelle le processus non-commutatif $(X_n, \ Y_n, \ Z_n)_{n \leq N}$ une marche de Bernoulli quantique.

Afin de mieux comprendre comment les trois processus X, Y, Z sont reliés, nous allons examiner plus en détails leur action sur l'espace $E^{\otimes N}$. Pour cela, introduisons les opérateurs

$$A_n^+ = \frac{1}{2}(X_n - iY_n), \ A_n^- = \frac{1}{2}(X_n + iY_n), \ B_n = \frac{1}{2}(nI - Z_n) \tag{8}$$

On a

$$A_n^+ = \sum_{k=0}^{n} \delta^{+(k)} \quad A_n^- = \sum_{k=0}^{n} \delta^{-(k)} \quad B_n = \sum_{k=0}^{n} \delta^{+(k)} \delta^{-(k)} \tag{9}$$

L'espace $E^{\otimes N}$ est muni d'une base orthonormale naturelle $(e_U)_{U \subset \{1,\dots,N\}}$ avec $e_U = e_{i_1} \otimes \dots \otimes e_{i_N}$ où $i_k = 1 \Leftrightarrow k \in U$ et $i_k = 0 \Leftrightarrow k \notin U$.

Dans cette base les opérateurs A_n^-, A_n^+, B_n sont donnés par les formules suivantes

$$A_n^+ e_U = \sum_{k \notin U, k \leq n} e_{U \cup \{k\}}$$

$$A_n^- e_U = \sum_{k \in U, k \leq n} e_{U \setminus \{k\}} \tag{10}$$

$$B_n e_U = |U \cap \{1, \dots, n\}| e_U$$

Considérons le sous-espace de $E^{\otimes N}$ engendré par les vecteurs orthogonaux ε_j^n définis par

$$\varepsilon_j^n = (C_n^j)^{-\frac{1}{2}} \sum_{|U \cap \{1,\dots,n\}|=j} e_U \text{ pour } j = 0, 1, \dots n \tag{11}$$

Ce sous-espace est laissé invariant par les opérateur A_n^+, A_n^-, B_n et on a

$$A_n^+ \varepsilon_j^n = \sqrt{(j+1)(n-j)}\varepsilon_{j+1}^n \qquad (12)$$

$$A_n^- \varepsilon_j^n = \sqrt{j(n-j+1)}\varepsilon_{j-1}^n \qquad (13)$$

$$B_n \varepsilon_j^n = j\varepsilon_j^n \qquad (14)$$

$$Z_n \varepsilon_j^n = (n-2j)\varepsilon_j^n \qquad (15)$$

pour $j = 0, 1, \ldots, n$ en posant $\varepsilon_{n+1}^n = \varepsilon_{-1}^n = 0$.

1.5 Le processus de spin

On a vu que la conjugaison par une matrice unitaire induit une rotation sur l'espace V. De même, si U est une matrice unitaire, l'opérateur $U^{\otimes N}$ agissant par conjugaison produit une rotation sur l'espace vectoriel réel engendré par les opérateurs X_n, Y_n, Z_n, pour $n \leq N$ et laisse invariantes les relations de commutation (3) entre ces opérateurs. Cela suggère d'étudier la "norme" de la marche de Bernoulli quantique, définie comme étant la racine carrée de $X_n^2 + Y_n^2 + Z_n^2$. Pour des raisons qui apparaitront plus bas, il est plus judicieux de considérer la racine carrée de $X_n^2 + Y_n^2 + Z_n^2 + I$.

DÉFINITION 7 – *On définit les variables* Σ_n, $0 \leq n \leq N$ *par les conditions*
$$\Sigma_n \text{ est positif}$$

$$\Sigma_n^2 = X_n^2 + Y_n^2 + Z_n^2 + I = 2(A_n^+ A_n^- + A_n^- A_n^+) + (2B_n - n)^2 + I \qquad (16)$$

Le processus $(\Sigma_n)_{n \leq N}$ *est appelé processus de spin.*

PROPOSITION 1 – *Si* $n \leq m$ *on a*

$$0 = [\Sigma_n, X_m] = [\Sigma_n, Y_m] = [\Sigma_n, Z_m] = [\Sigma_n, \Sigma_m]$$

Démonstration
Des relations (7) on tire pour $n \leq m$: $[Y_n^2, X_m] = -2i(Y_n Z_n + Z_n Y_n)$ et $[Z_n^2, X_m] = 2i(Y_n Z_n + Z_n Y_n)$ d'où $[\Sigma_n^2, X_m] = 0$ et $[\Sigma_n, X_m] = 0$. Les autres relations en découlent par permutation circulaire de X_m, Y_m, Z_m.

Le processus de spin est un processus classique, on peut donc se poser le problème de déterminer sa loi. C'est ce que nous allons faire dans la suite du paragraphe, et pour cela on va déterminer une base de vecteurs propres communs à tous les opérateurs $(\Sigma_n)_{1 \leq n \leq N}$.

Remarquons tout de suite que l'on a, d'après (12), (13), (14) et (16)

$$\Sigma_n \varepsilon_j^n = (n+1)\varepsilon_j^n$$

Les vecteurs ε_j^n sont donc des vecteurs propres de Σ_n. Nous allons généraliser ce résultat dans la proposition suivante.

PROPOSITION 2 – *Soit* $J = (j_1, \ldots, j_N)$ *une suite d'entiers telle que*
a) $j_1 = 2$
b) $j_i \geq 1$ *pour tout* $i \leq N$.
c) $j_{i+1} - j_i \in \{+1, -1\}$ $\forall i < N$

Alors il existe un sous-espace H_J de $E^{\otimes N}$ de dimension $j_N = l + 1$, espace propre commun à $\Sigma_1, \ldots, \Sigma_N$ de valeurs propres correspondantes $(j_1, \ldots, j_N)$, avec une base orthonormale $(\phi_0^J, \phi_1^J, \ldots, \phi_l^J)$ telle que

$$A_N^+ \phi_j^J = \sqrt{(j+1)(l-j)}\, \phi_{j+1}^J$$
$$A_N^- \phi_j^J = \sqrt{j(l-j+1)}\, \phi_{j-1}^J \qquad (17)$$
$$Z_N \phi_j^J = (2j-l)\, \phi_j^J$$

(avec $\phi_{l+1}^J = \phi_{-1}^J = 0$). De plus, les espace H_J sont orthogonaux et $E^{\otimes N} = \oplus_J H_J$.

Démonstration

On le démontre par récurrence sur N.

C'est vrai pour N=1, avec $J = (2)$, $H_J = E$ et $e_0 = \phi_0^J$, $e_1 = \phi_1^J$.

Soit $J' = (j_1, \ldots, j_{N+1})$ une suite de longueur N+1 satisfaisant les hypothèses a) et b), et $J = (j_1, \ldots, j_N)$. Par hypothèse, l'espace $H_J \subset E^{\otimes N}$ a une base $\phi_0^J, \ldots \phi_l^J$ avec $j_N = l + 1$ vérifiant (17). Le sous-espace $H_J \otimes E \subset E^{\otimes(N+1)}$ est un sous-espace propre commun à $\Sigma_1, \ldots, \Sigma_N$ sur lequel ces opérateurs admettent les valeurs propres $(j_1, \ldots j_N)$. Nous allons le décomposer sous l'action de Σ_{N+1}.

Posons

$$\eta_j = \sqrt{\frac{l-j+1}{l+1}}\, \phi_j^J \otimes e_0 + \sqrt{\frac{j}{l+1}}\, \phi_{j-1}^J \otimes e_1 \quad 0 \leq j \leq l+1$$

$$\xi_j = \sqrt{\frac{j+1}{l+1}}\, \phi_j^J \otimes e_0 - \sqrt{\frac{l-j}{l+1}}\, \phi_{j-1}^J \otimes e_1 \quad 0 \leq j \leq l-1$$

Les vecteurs η_j, ξ_j forment une base orthonormale de $H_J \otimes E$.

Comme $A_{N+1}^+ = A_N^+ + \delta^{+(N+1)}$ et $A_{N+1}^- = A_N^- + \delta^{-(N+1)}$ un calcul simple utilisant les formules pour A_N^+, A_N^- et Z_N donne

$$A_{N+1}^+ \eta_j = \sqrt{(j+1)(l+1-j)}\, \eta_{j+1}$$
$$A_{N+1}^- \eta_j = \sqrt{j(l-j+2)}\, \eta_{j-1}$$
$$Z_{N+1} \eta_j = (l+1-2j)\, \eta_j$$
$$\Sigma_{N+1} \eta_j = (l+1)\, \eta_j$$

pour $0 \leq j \leq l+1$ et

$$A_{N+1}^+ \xi_j = \sqrt{(j+1)(l-1-j)}\, \xi_{j+1}$$
$$A_{N+1}^- \xi_j = \sqrt{j(l-j)}\, \xi_{j-1}$$
$$Z_{N+1} \xi_j = (l-1-2j)\, \xi_j$$
$$\Sigma_{N+1} \xi_j = (l-1)\, \xi_j$$

pour $0 \leq j \leq l-1$, en posant $\xi_l = \eta_{l+2} = \xi_{-1} = \eta_{-1} = 0$. On en déduit qu'en posant

$H_{J'} = vect(\eta_j)_{0 \leq j \leq l+1}, \phi_j^{J'} = \eta_j$ si $j_{N+1} = J_N + 1$

$H_{J'} = vect(\xi_j)_{0 \leq j \leq l-1}, \phi_j^{J'} = \xi_j$ si $j_{N+1} = J_N - 1$

alors $H_{J'}$ est un sous-espace propre commun aux Σ_j, de suite de valeurs propres J', et on a le résultat de la proposition au rang $N + 1$.

Nous pouvons maintenant calculer la loi du processus Σ.

THÉORÈME 1 – Si $p = 1$ la loi de Σ est celle d'un processus déterministe, la trajectoire $(2, 3, \ldots, N + 1)$ ayant une probabilité 1.

Si $p = \frac{1}{2}$, la loi de Σ est celle d'une chaîne de Markov sur $\mathbb{N}^*$ de probabilités de transition:

$$p(n, n + 1) = \frac{n + 1}{2n} \quad p(n, n - 1) = \frac{n - 1}{2n} \tag{18}$$

Démonstration

Tout d'abord, si $p = 1$, l'état $\Sigma^{\otimes N}$ est l'état pur associé au vecteur $e_0^{\otimes N}$, donc c'est un vecteur propre commun à $\Sigma_1, \ldots, \Sigma_N$, de valeurs propres $2, 3, \ldots, N + 1$, et la loi de Σ est celle d'un processus déterministe, qui met la probabilité 1 sur la trajectoire $(2, 3, \ldots, N + 1)$.

Si $p = \frac{1}{2}$, la probabilité d'une trajectoire $J = (j_1, \ldots, j_N)$ est égale à $\frac{1}{2^N} tr(\Pi_J)$ où Π_J est le projecteur orthogonal sur le sous espace H_J. Comme $dim(H_J) = j_N$ cette probabilité est $\frac{j_N}{2^N}$, et il est facile d'en déduire que la loi de Σ est celle d'une chaîne de Markov sur $\mathbb{N}^*$ dont les probabilités de transition sont données par celles du théorème.

REMARQUES– Cette dernière chaîne de Markov est bien connue en probabilités classiques, il s'agit d'un analogue discret du processus de Bessel de dimension 3 (cf J.Pitman [68]). En fait, si $(B_t)_{t \geq 0}$ est un mouvement brownien de dimension 3 issu de 0, et les instants T_n sont les instants successifs de passage de $|B_t|$ par des valeurs entières, (i.e. $T_0 = 0$ et $T_{n+1} = \inf\{t \geq T_n \mid |B_t| - |B_{T_n}| = \pm 1\}$) alors le processus $|B_{T_n}|_{n \geq 0}$ est une chaîne de Markov sur $\mathbb{N}$ dont les probabilités de transition sont données par la formule (18)

Si U, V, W sont trois marches de Bernoulli symétriques indépendantes, leur norme n'est pas un processus de Markov, et n'est pas à valeurs entières. Les propriétés du processus Σ reflètent l'invariance par rotation du triplet X, Y, Z (ou plus fondamentalement, des relations de commutation (3)). La marche de Bernoulli quantique peut être considérée comme une approximation discrète du mouvement brownien de dimension 3, invariante par rotation.

Nous verrons au chapitre 6 que l'on peut interpréter ce processus comme un processus de Markov à valeurs dans un espace non-commutatif. Cela proviendra de résultats de la théorie des représentations des groupes. Les calculs de la proposition 2 qui proviennent de la théorie des représentations de l'algèbre de Lie $sl(2, \mathbb{C})$ en sont déjà un exemple (comparer à [78]).

Exercice: Calculer la loi de $(\Sigma_n)_{n \geq 1}$ pour $p \neq 1, \frac{1}{2}$.

1.6 Un théorème limite pour la marche de Bernoulli quantique

Il y a essentiellement deux théorèmes limite pour l'addition de variables de Bernoulli indépendantes. Le premier est le théorème de Moivre-Laplace (cas particulier du théorème central limite), qui concerne l'addition de variables de Bernoulli symétriques et qui donne à la limite une loi normale. Le second est le théorème de convergence des lois binomiales de paramètres (N, α_N) vers la loi de Poisson de paramètre α lorsque $N \to +\infty$ et $N\alpha_N \to \alpha$. On peut réaliser la loi binomiale (N, α_N) comme loi d'une somme de N variables de Bernoulli indépendantes de même loi sur $\{0, 1\}$ donnée par $P(\{0\}) = 1 - \alpha_N$, $P(\{1\}) = \alpha_N$.

Nous avons vu au paragraphe 1.3 que pour $p = 1$, l'espace de probabilité non-commutatif fini (E, S) permet de réaliser toutes les lois de Bernoulli sur $\mathbb{R}$. Cela va nous permettre de donner un "théorème limite non-commutatif" qui englobe les deux résultats cités ci-dessus. Plus précisément, pour tout $\theta \in \mathbb{R}$ la variable $e^{i\theta}\delta^+ + e^{-i\theta}\delta^-$ suit une loi de Bernoulli symétrique sur $\{+1, -1\}$, donc d'après le théorème de Moivre-Laplace, dans l'état $S^{\otimes N}$, la loi de $e^{i\theta}\frac{A_N^+}{\sqrt{N}} + e^{-i\theta}\frac{A_N^-}{\sqrt{N}}$ converge vers la loi normale centrée réduite.

De même, on peut construire une variable de Bernoulli sur $\{0, 1\}$ en utilisant l'opérateur

$$(1 + \frac{|z|^2}{N})^{-1} \begin{pmatrix} \frac{|z|^2}{N} & \frac{\bar{z}}{\sqrt{N}} \\ \frac{z}{\sqrt{N}} & 1 \end{pmatrix} = (1 + \frac{|z|^2}{N})^{-1}((1 - \frac{|z|^2}{N})\delta^+\delta^- + \frac{z}{\sqrt{N}}\delta^+ + \frac{\bar{z}}{\sqrt{N}}\delta^- + \frac{|z|^2}{N}Id)$$

Sa loi est portée par $\{0, 1\}$, et $P(\{0\}) = \frac{1}{1 + \frac{|z|^2}{N}}$, $P(\{1\}) = \frac{\frac{|z|^2}{N}}{1 + \frac{|z|^2}{N}}$, par conséquent, dans l'état $S^{\otimes N}$, l'opérateur

$$(1 + \frac{|z|^2}{N})^{-1}((1 - \frac{|z|^2}{N})B_N + \frac{zA_N^+}{\sqrt{N}} + \frac{\bar{z}A_N^-}{\sqrt{N}} + |z|^2 Id)$$

suit une loi binomiale de paramètres $(N, \frac{\frac{|z|^2}{N}}{1 + \frac{|z|^2}{N}})$ qui converge vers la loi de Poisson de paramètre $|z|^2$ quand $N \to +\infty$.

Ces deux résultats classiques suggèrent d'étudier la convergence "en loi" du triplet d'opérateurs $(B_N, \frac{A_N^+}{\sqrt{N}}, \frac{A_N^-}{\sqrt{N}})$. Comme ces opérateurs ne sont pas auto-adjoints et ne commutent pas, nous ne pouvons pas considérer leur loi jointe, mais néanmoins nous pouvons calculer leurs "moments" c'est à dire les expressions de la forme $< P(B_N, \frac{A_N^+}{\sqrt{N}}, \frac{A_N^-}{\sqrt{N}})e_0^{\otimes N}, e_0^{\otimes N} >$ où P est un polynôme en trois indéterminées non-commutatives. Le théorème suivant montre que ces moments convergent quand $N \to +\infty$, vers les moments d'un triplet (a^0, a^+, a^-) d'opérateurs sur un espace de dimension infinie.

Les opérateurs (a^0, a^+, a^-) vont jouer un rôle fondamental par la suite, et nous les étudierons en détail au chapitre 3, après avoir revu les notions nécessaires de théorie spectrale au chapitre 2.

Soit H_0 un espace vectoriel complexe, pré-hilbertien de dimension infinie avec une base orthonormale, $(\varepsilon_0 = \Omega, \varepsilon_1, \ldots, \varepsilon_n, \ldots)$. On définit des applications linéaires (a^+, a^-, a^0) sur H_0 par les formules

$$\begin{aligned} a^+\varepsilon_j &= \sqrt{j+1}\varepsilon_{j+1} \\ a^-\varepsilon_j &= \sqrt{j}\varepsilon_{j-1} \\ a^0\varepsilon_j &= j\varepsilon_j \end{aligned} \tag{19}$$

pour $j \geq 0$, avec $\varepsilon_{-1} = 0$.

THÉORÈME 2 – Pour tout polynome P en trois indéterminées non-commutatives, on a

$$\lim_{N \to \infty} < P(B_N, \frac{A_N^+}{\sqrt{N}}, \frac{A_N^-}{\sqrt{N}})e_0^{\otimes N}, e_0^{\otimes N} > \; = \; < P(a^0, a^+, a^-)\Omega, \Omega >$$

Démonstration

Il suffit de le montrer pour un monome de degré d, pour tout entier $d \geq 0$. Le vecteur $e_0^{\otimes N}$ est égal à ε_0^N d'après (11). Le sous-espace de $E^{\otimes N}$ engendré par les vecteurs ε_j^N est stable par application des opérateurs B_N, A_N^+, A_N^-, par conséquent dans le calcul de $< P(B_N, \frac{A_N^+}{\sqrt{N}}, \frac{A_N^-}{\sqrt{N}})\varepsilon_0^N, \varepsilon_0^N >$ seules interviennent les valeurs des opérateurs $B_N, \frac{A_N^+}{\sqrt{N}}, \frac{A_N^-}{\sqrt{N}}$ évalués sur ε_j^N pour $j \leq d$. Or d'après (12)(13)(14) on a

$$\frac{A_N^+}{\sqrt{N}}\varepsilon_j^N = \sqrt{\frac{(j+1)(N-j)}{N}}\,\varepsilon_{j+1}^N$$

$$\frac{A_N^-}{\sqrt{N}}\varepsilon_j^N = \sqrt{\frac{j(N-j+1)}{N}}\,\varepsilon_{j-1}^N$$

$$B_N\varepsilon_j^N = j\varepsilon_j^N$$

le théorème suit en comparant ces formules avec (19) quand $N \to \infty$.

Comme les opérateurs $(B_N, \frac{A_N^+}{\sqrt{N}}, \frac{A_N^-}{\sqrt{N}})$ ne commutent pas, ils n'ont pas de loi jointe au sens classique et ce théorème est donc purement non-commutatif. Cependant, si on considère des polynômes $P(B_N, \frac{A_N^+}{\sqrt{N}}, \frac{A_N^-}{\sqrt{N}})$ qui s'expriment sous la forme d'un polynôme en $(e^{i\theta}\frac{A_N^+}{\sqrt{N}} + e^{-i\theta}\frac{A_N^-}{\sqrt{N}})$ ou en $((1+\frac{|z|^2}{N})^{-1}((1-\frac{|z|^2}{N})B_N + \frac{zA_N^+}{\sqrt{N}} + \frac{\bar{z}A_N^-}{\sqrt{N}} + |z|^2 Id)$ les théorèmes limite cités plus haut montrent qu'il y a convergence de ces moments vers les moments correspondants des lois de Gauss ou de Poisson. Cela suggère que les opérateurs $e^{i\theta}a^+ + e^{-i\theta}a^-$ "sont" des variables normales, tandis que les opérateurs $a^0 + za^+ + \bar{z}a^- + |z|^2$ "sont" des variables de Poisson. C'est ce que nous vérifierons dans le chapitre 3, après avoir vu les notions nécessaires de théorie spectrale au chapitre 2.

Chapitre 2

Théorie spectrale

2.1 Quelques définitions

Après les espaces de probabilité non-commutatifs finis nous allons passer au cadre plus général des espaces de Hilbert de dimension infinie, qui nous permettra de considérer des variables non-commutatives dont les lois ne seront pas à support fini. Pour cela, nous allons devoir faire des rappels d'analyse fonctionnelle, plus précisément de théorie des opérateurs auto-adjoints. Cette théorie classique, due entre autres à von Neumann et Stone est exposée par exemple dans les livres de Reed et Simon [70], Rudin [72], Halmos [39].

Dans la suite, H désigne un espace de Hilbert complexe séparable.
Commençons par quelques définitions.

DÉFINITION 8 – *Un* opérateur T *sur* H *est la donnée d'un couple* $(D(T), T)$ *où* $D(T)$, *le domaine de l'opérateur, est un sous-espace vectoriel de* H *et* T *une application linéaire de* $D(T)$ *dans* H.
L'opérateur T *est dit borné si* $D(T) = H$ *et* $\|T\| = \sup_{x \in H, x \neq 0} \frac{|Tx|}{|x|} < +\infty$.

Dans la suite, quand on parlera d'un opérateur, son domaine sera souvent sous-entendu. Les opérateurs bornés sur H forment une algèbre notée $\mathcal{B}(H)$, qui est une algèbre de Banach pour la norme $\| \|$. Il y a d'autres topologies intéressantes sur $\mathcal{B}(H)$, citons entre autres la topologie forte, qui est celle de la convergence simple sur H, et qui est (malgré son nom) plus faible que celle de la norme.

EXEMPLE FONDAMENTAL– Soient $(\Omega, \mathcal{F}, P)$ un espace de probabilités, $H = L^2_{\mathbb{C}}(\Omega, \mathcal{F}, P)$, et $X : \Omega \to \mathbb{C}$ une variable aléatoire.
Posons $D(T) = \{f \in L^2_{\mathbb{C}}(\Omega, \mathcal{F}, P) | Xf \in L^2_{\mathbb{C}}(\Omega, \mathcal{F}, P)\}$, et $T_X(f) = Xf$, pour $f \in D(T)$, alors T_X est un opérateur. L'opérateur T_X ne dépend que de la classe de X modulo les ensembles de mesure nulle, et il est borné si et seulement si $X \in L^\infty(\Omega, \mathcal{F}, P)$. Cet exemple est une généralisation immédiate de l'exemple du début du chapitre 1.

DÉFINITION 9 – *Le graphe de l'opérateur* T *est l'ensemble*

$$\Gamma(T) = \{(x, Tx) \in H \times H | x \in D(T)\}$$

T *est dit fermé si* $\Gamma(T)$ *est fermé dans* $H \times H$.
Un opérateur S *est une extension de* T *(on note* $T \subset S$*) si* $\Gamma(T) \subset \Gamma(S)$ *(ou encore* $D(T) \subset D(S)$ *et* $T(x) = S(x)$, $\forall x \in D(T)$*).*
Un opérateur T *est dit fermable s'il possède une extension fermée, et sa fermeture est alors l'opérateur dont le graphe est* $\overline{\Gamma(T)}$.

DÉFINITION 10 – *On suppose* $D(T)$ *dense dans* H. *L'adjoint de* T *est l'opérateur* T^* *de domaine*

$D(T^*) = \{y \in H | x \mapsto < Tx, y >$ *est continue sur* $D(T)\}$ *tel que pour* $y \in D(T^*)$ *on a* $< Tx, y >=< x, T^*y > \forall x \in D(T)$ ($D(T)$ *étant dense dans* H *cette dernière égalité détermine* T^*y).

Un opérateur de la forme T^* *est toujours fermé.*

T *est dit symétrique si* $T \subset T^*$ *(ou encore* $< Tx, y >=< x, Ty > \forall x, y \in D(T)$*)*

L'opérateur T *est dit auto-adjoint si* $T = T^*$, *unitaire s'il est borné, inversible, et* $T^* = T^{-1}$.

Revenons à l'exemple fondamental: si X est une variable aléatoire réelle, l'opérateur T_X associé est auto-adjoint. Réciproquement, le théorème spectral (énoncé ci-dessous) permet de réaliser tout opérateur auto-adjoint comme un opérateur de multiplication sur un espace L^2.

DÉFINITION 11 – *Le spectre de l'opérateur* T *est l'ensemble*

$$\sigma(T) = \{\lambda \in \mathbb{C} \mid T - \lambda Id \text{ n'a pas d'inverse borné}\}$$

Le spectre d'un opérateur auto-adjoint est un sous-ensemble fermé de $\mathbb{R}$.

L'opérateur auto-adjoint T *est dit positif si son spectre est inclus dans* $\mathbb{R}_+$.

Les valeurs propres de T *sont les* $\lambda \in \mathbb{C}$ *tels que* $ker(T - \lambda Id) \neq \{0\}$.

DÉFINITION 12 – *Un projecteur est un opérateur auto-adjoint borné* P *tel que* $P^2 = P$. *C'est l'opérateur de projection orthogonale sur son image.*

2.2 Résolutions de l'identité et théorème spectral

Soient $(A, \mathcal{A})$ un espace mesurable et H un espace de Hilbert.

DÉFINITION 13 – *Une résolution de l'identité sur* $(A, \mathcal{A})$ *est une famille* $(E_X)_{X \in \mathcal{A}}$ *de projecteurs orthogonaux de* H *telle que*

i) $E_A = id$, $E_\emptyset = 0$

ii) $\forall X_1, X_2 \in \mathcal{A}$ $E_{X_1} E_{X_2} = E_{X_1 \cap X_2}$ $\qquad\qquad$ (20)

iii) Pour toute famille dénombrable $(X_j)_{j \in J}$

$\qquad$ *telle que* $X_j \cap X_k = \emptyset$ *si* $k \neq j$ *et tout* $x \in H$ *on a*

$$E_{\cup_{j \in J} X_j}(x) = \sum_{j \in J} E_{X_j}(x)$$

Une résolution de l'identité est portée par $X \in \mathcal{A}$ *si* $Y \cap X = \emptyset \Rightarrow E_Y = 0$.

Soit $f : (A, \mathcal{A}) \to (B, \mathcal{B})$ une application mesurable, on définit une résolution de l'identité $f(E)$ sur $(B, \mathcal{B})$ en posant $f(E)_Y = E_{f^{-1}(Y)}$.

Soit E une résolution de l'identité sur $(A, \mathcal{A})$, pour tout couple de vecteurs $x, y \in H$ d'après i) ii) et iii), on définit une mesure complexe bornée $E_{x,y}$ sur $(A, \mathcal{A})$ par $E_{x,y}(B) = < E_B(x), y >$.

On notera $E_x = E_{x,x}$, si $x \in H$.

Soit E une résolution de l'identité sur $\mathbb{R}$, posons

$$D(T) = \{x \in H | \int_{\mathbb{R}} \lambda^2 \, dE_x(\lambda) < +\infty\} \qquad\qquad (21)$$

$$< Tx,y >= \int_{\mathbb{R}} \lambda dE_{x,y}(\lambda) \tag{22}$$

Ces formules définissent ainsi un opérateur auto-adjoint T. Réciproquement, le théorème spectral montre qu'à tout opérateur auto-adjoint T on peut associer une résolution de l'identité pour laquelle T est définie par (21) et (22).

THÉORÈME 3 – *Soit T un opérateur auto-adjoint sur H, il existe une unique résolution de l'identité E sur $\mathbb{R}$, portée par $\sigma(T)$, telle que*

$$D(T) = \{x \in H | \int_{\mathbb{R}} \lambda^2 \, dE_x(\lambda) < +\infty\}$$

$$\forall x,y \in D(T) \quad < Tx,y >= \int_{\mathbb{R}} \lambda dE_{x,y}(\lambda)$$

Dans le cas où H est de dimension finie, la résolution de l'identité est facile à décrire. Le spectre de l'opérateur T est l'ensemble de ses valeur propres, et si λ est une valeur propre de T, $E_{\{\lambda\}}$ est le projecteur orthogonal sur le sous-espace propre correspondant.

On déduit du théorème spectral le résultat suivant. Si T est un opérateur auto-adjoint sur H, il existe un espace de probabilités $(\Omega, \mathcal{F}, P)$, une isométrie $\iota : H \to L^2(\Omega, \mathcal{F}, P)$ et une variable aléatoire réelle X telle que l'opérateur de multiplication par X soit égal à $\iota \circ T \circ \iota^*$ (voir par exemple Reed et Simon [70]).

Une autre conséquence du théorème spectral est la possibilité d'un calcul borélien sur les opérateurs auto-adjoints. Soient T un opérateur auto-adjoint de résolution de l'identité E, et f une fonction borélienne sur $\sigma(T)$, on peut définir un opérateur auto-adjoint $f(T)$ par sa résolution de l'identité $f(E)$. En particulier, si f est bornée, $f(T)$ est borné. Ce calcul fonctionnel prolonge le calcul polynomial usuel sur les opérateurs bornés et il respecte les opérations algèbriques:

$$(f + g)(T) = f(T) + g(T), fg(T) = f(T)g(T), (\lambda f)(T) = \lambda \, f(T)$$

pour des fonctions f et g bornées, et $\lambda \in \mathbb{R}$.
De plus, si $f_n \to f$ simplement, et les f_n sont uniformément bornées, alors $f_n(T) \to f(T)$ fortement (cette dernière propriété est une conséquence du théorème de convergence dominée de Lebesgue).

DÉFINITION 14 – *Deux résolutions de l'identité E et E' sur $(A, \mathcal{A})$ et $(A', \mathcal{A}')$ commutent si*

$$\forall X \in \mathcal{A}, Y \in \mathcal{A}' \quad E_X E'_Y = E'_Y E_X$$

PROPOSITION 3 – *Soient $(A_j, \mathcal{A}_j)_{j \in J}$ des espaces lusiniens. Soit $(E^j)_{j \in J}$ une famille de résolutions de l'identité sur $(A_j, \mathcal{A}_j)_{j \in J}$, commutant deux à deux, il existe une unique résolution de l'identité E sur $(\prod_{j \in J} A_j, \otimes_{j \in J} \mathcal{A}_j)$ telle que pour tout ensemble cylindrique $\prod_{j \in J} X_j$ (où $X_j = A_j$ sauf pour un nombre fini d'indices) on ait*

$$E_{\prod_{j \in J} X_j} = \prod_{j \in J} E^j_{X_j}$$

(le produit dans le membre de droite est bien défini car $E_{X_j}^j = Id$ sauf pour un nombre fini de $j \in J$ et ces projecteurs commutent).

Démonstration

Soit $\phi \in H$, posons $\mu_{\phi,\phi}(\prod_{j \in J} X_j) = \prod_{j \in J} < E_{X_j}^j(\phi), \phi >$ pour un ensemble cylindrique $\prod_{j \in J} X_j$. On définit ainsi une application additive sur la sous-algèbre de Boole de $\otimes_{j \in J} \mathcal{A}_j$ engendrée par les ensembles cylindriques $\prod_{j \in J} X_j$. Soit U_n une suite décroissante dans cette algèbre de Boole telle que $\mu_{\phi,\phi}(U_n) \geq \varepsilon > 0$ pour tout n. Pour tout n il existe un compact $V_n \subset U_n$ tel que $\mu_{\phi,\phi}(V_n) \geq \varepsilon(1 - 3^{-n})$. Il s'ensuit que $\cap_{n \leq N} V_n$ est une suite décroissante de compacts non vides donc $\cap_{n \geq 0} V_n \neq \emptyset$ et $\cap_{n \geq 0} U_n \neq \emptyset$. D'après le théorème de Carathéodory, $\mu_{\phi,\phi}$ se prolonge en une mesure positive sur $\otimes_{j \in J} \mathcal{A}_j$.

La résolution de l'identité E sur $\otimes_{j \in J} \mathcal{A}_j$ est obtenue en posant $< E_X(\phi), \phi >= \mu_{\phi,\phi}(X)$, ce qui détermine E_X de façon unique par polarisation.

On dit que deux opérateurs auto-adjoints commutent si leurs résolutions de l'identité commutent (pour des opérateurs bornés, cela revient au même que la commutation usuelle).

On peut alors définir un calcul fonctionnel mesurable pour une famille $(T_j)_{j \in J}$ d'opérateurs auto-adjoints qui commutent en posant que $f((T_j))$ est l'opérateur auto-adjoint dont la résolution de l'identité est $f(E)$, où f est une fonction mesurable sur $\prod_{j \in J} \sigma(X_j)$ et E la résolution de l'identité sur $\prod_{j \in J} \sigma(X_j)$ obtenue en appliquent la proposition précédente aux résolutions des opérateurs X_j. Ce calcul vérifie les mêmes propriétés que celui décrit plus haut pour un seul opérateur.

Comme dans le cas d'un seul opérateur, on montre qu'il existe un espace de probabilité $(\Omega, \mathcal{F}, P)$, une isométrie: $\iota : H \to L^2(\Omega, \mathcal{F}, P)$ et des variable aléatoire réelle X_j telles que les opérateur de multiplication par X_j soient $\iota \circ T_j \circ \iota^*$.

2.3 Opérateurs à trace et espaces de probabilité non-commutatifs

Soit $\phi \in H$ un vecteur de norme 1, et E une résolution de l'identité sur $(A, \mathcal{A})$, on obtient une loi de probabilité sur $(A, \mathcal{A})$ en posant $\mu(X) =< E_X(\phi), \phi >$. Lorsque E est la résolution de l'identité d'un opérateur auto-adjoint T, la mesure μ est une loi de probabilité sur $\sigma(T)$. De même, si E est la résolution de l'identité associée à une famille d'opérateurs qui commutent, on a ainsi une loi sur le produit des spectres de ces opérateurs.

DÉFINITION 15 – *La forme linéaire sur $\mathcal{B}(H)$ définie par $T \to< T\phi, \phi >$ est appelée l'état pur associé à ϕ.*

Comme dans le cas de la dimension finie, on considèrera des états obtenus comme combinaison convexe d'états purs. Ces combinaisons sont décrites au moyen de la notion d'opérateur à trace.

DÉFINITION 16 – *Soit T un opérateur borné, on dit que T est à trace si $\sum_{n=0}^{+\infty} | < Te_n, e_n > | < +\infty$ pour une base orthonormée $(e_n)_{n \in \mathbb{N}}$ de H. (Cette propriété ne dépend pas de la base choisie). La trace de T est la quantité $tr(T) = \sum_{n=0}^{+\infty} < Te_n, e_n >$, qui ne dépend pas de la base utilisée pour la calculer.*

Les opérateurs à trace forment un idéal bilatère de l'algèbre $\mathcal{B}(H)$.

DÉFINITION 17 – *Si S est un opérateur à trace, positif, de trace 1, la forme linéaire sur $\mathcal{B}(H)$ définie par $T \mapsto tr(TS)$ est appelée l'état associé à S (par abus de langage on appelle aussi parfois état l'opérateur S lui-même).*

Soit π_ϕ le projecteur orthogonal sur le sous-espace engendré par un vecteur ϕ de norme 1, alors $tr(\pi_\phi T) =< T\phi, \phi >$ est l'état pur associé à ϕ. Comme en dimension finie, l'ensemble des états est l'enveloppe convexe faiblement fermée, dans le dual de $\mathcal{B}(H)$, de l'ensemble des états purs, qui en sont les points extrémaux.

On peut maintenant définir la notion d'espace de probabilité non-commutatif.

DÉFINITION 18 – *Un espace de probabilité non-commutatif est la donnée d'un couple (H, S) où H est un espace de Hilbert et S un opérateur positif à trace, de trace 1.*

DÉFINITION 19 – *Une variable aléatoire non-commutative sur l'espace (H, S) est un opérateur auto-adjoint sur H.*
Une famille de $(T_j)_{j \in J}$ d'opérateurs sur H est appelée un processus non-commutatif, et si ces opérateurs sont auto-adjoints et commutent, le processus est dit classique.

Soit $(T_j)_{j \in J}$ un processus classique, et E la résolution de l'identité sur $\prod_{j \in J} \sigma(T_j)$ associée, la loi jointe des T_j dans l'état S, est la mesure sur $\prod_{j \in J} \sigma(T_j)$
$X \mapsto tr(E_X S)$, c'est la loi du processus classique $(T_j)j \in J$.

Les définitions que nous venons de voir sont les généralisations en dimension infinie des définitions du chapitre 1. Une famille de variables aléatoires réelles au sens usuel peut s'interpréter comme une famille d'opérateurs auto-adjoints sur leur espace L^2, ce qui fait que la théorie des probabilités non-commutative contient la théorie usuelle. Chaque fois que l'on rencontre, en probabilités non-commutatives un processus classique, on est en situation d'appliquer les résultat usuels de probabilités. La situation se complique (mais devient aussi plus intéressante) lorsque l'on considère *simultanément* des variables qui ne commutent pas entre elles. Une des conséquences est que l'on ne peut pas parler des trajectoires d'un processus non-commutatif, tout ce qu'on en connait sont des propriétés en loi. En un sens, on est ramené à la situation des probabilités classiques avant Kolmogorov, lorsqu'on ne possédait pas d'espace de probabilité, et que l'on ne savait parler des processus qu'en loi.

2.4 Théorèmes de Stone et de Nelson

Dans la suite nous aurons à travailler avec des opérateurs symétriques définis sur un domaine dense d'un espace de Hilbert H. Afin de leur appliquer la théorie spectrale, il faudra leur trouver une extension auto-adjointe. Le but de ce paragraphe est de fournir des critères pour l'existence et l'unicité d'une telle extension.

DÉFINITION 20 – *Un opérateur symétrique est dit essentiellement auto-adjoint s'il admet une seule extension auto-adjointe (ou, ce qui est équivalent, si sa fermeture est auto-adjointe).*

DÉFINITION 21 – *Un groupe à un paramètre fortement continu d'opérateurs unitaires est une famille $(U_t)_{t \in \mathbb{R}}$ d'opérateurs unitaires telle que $\forall s, t \in \mathbb{R}$ $U_t U_s = U_{t+s}$, et l'application $t \mapsto U_t x$ est continue pour tout $x \in H$.*

THÉORÈME 4 – *Soit* $(U_t)_{t \in \mathbb{R}}$ *un groupe fortement continu d'opérateurs unitaires, et*

$$D(A) = \{\phi \in H \,|\, \lim_{t \to 0} \frac{1}{i} \frac{U_t(\phi) - \phi}{t} \text{ existe}\}$$

alors $D(A)$ *est dense dans* H, *et l'opérateur* A *défini sur le domaine* $D(A)$ *par* $A\phi = \lim_{t \to 0} \frac{1}{i} \frac{U_t(\phi) - \phi}{t}$ *est un opérateur auto-adjoint, de plus on a* $U_t = e^{itA}$.

Ce théorème est dû à Stone (cf [70] ou [72]).

DÉFINITION 22 – *Soit* A *un opérateur,* $\phi \in D(A)$ *est dit analytique pour* A *si* $A\phi \in D(A)$, $A^2\phi \in D(A)$, ... $A^n\phi \in D(A)$ *pour tout entier* n, *et il existe* $t \in \mathbb{R}_+^*$ *tel que* $\sum_{n=0}^{+\infty} \frac{t^n}{n!} |A^n\phi| < +\infty$.

THÉORÈME 5 – *Soit* A *un opérateur symétrique, possédant un sous-espace dense de vecteurs analytiques, alors* A *est essentiellement auto-adjoint.*

Ce théorème est dû à Nelson ([55] voir Reed et Simon [70]).

Chapitre 3

Variables de Gauss et de Poisson et relations de commutation

3.1 Relations de commutation d'Heisenberg

Après ces points de théorie spectrale nous allons reprendre le fil, en étudiant les opérateurs introduits à la fin du chapitre 1 dans le théorème limite pour les marches de Bernoulli quantiques.

Soit H un espace de Hilbert complexe muni d'une base orthonormale $(e_n)_{n\in\mathbb{N}}$. On notera parfois Ω le vecteur e_0, et on appellera "état vide" l'état pur associé. On considère les opérateurs a^0, a^+, a^- de domaine $D = vect(e_n)_{n\in\mathbb{N}}$, définis par

$$
\begin{aligned}
a^0 e_n &= n\, e_n \\
a^+ e_n &= \sqrt{n+1}\, e_{n+1} \\
a^- e_n &= \sqrt{n}\, e_{n-1}
\end{aligned}
\tag{23}
$$

(avec la convention $e_{-1} = 0$).

Ce chapitre est consacré à l'étude approfondie de ces opérateurs. Tout d'abord, d'un point de vue analytique, nous allons montrer que les combinaisons linéaires symétriques de ces opérateurs sont essentiellement auto-adjointes, puis nous calculerons les lois dans l'état vide de ces variables aléatoires. On vérifiera que l'on obtient bien des lois normales et des lois de Poisson comme le laisse présager le théorème limite (2). Le triplet (a^0, a^+, a^-) joue en probabilités quantiques le rôle qui est tenu par les variables aléatoires de Gauss et de Poisson en probabilités classiques. Ce qui est remarquable ici, ce sont les relations algébriques simples entre ces opérateurs.

Les opérateurs a^0, a^+, a^- laissent stable le domaine D, ce qui permet de les composer et de donner les relations suivantes

$$
a^0 = a^+ a^- \tag{24}
$$
$$
[a^+, a^-] = -Id \tag{25}
$$
$$
[a^0, a^+] = a^+ \tag{26}
$$
$$
[a^0, a^-] = -a^- \tag{27}
$$

Les opérateurs a^+ et a^- sont formellement adjoints l'un de l'autre, c'est-à-dire que

$$
< a^+ x, y > = < x, a^- y > \quad \forall x, y \in D \tag{28}
$$

Ceci entraine que $a^+ \subset (a^-)^*$ et $a^- \subset (a^+)^*$, et en particulier les opérateurs a^+ et a^- sont fermables.

Les combinaisons linéaires

$$
Q = a^+ + a^- \text{ et } P = \frac{1}{i}(a^+ - a^-)
$$

sont des opérateurs symétriques. En terme de ces opérateurs, les relations (25) s'écrivent

$$a^0 = \frac{1}{2}(P^2 + Q^2) \tag{29}$$

$$[P, Q] = 2iId \tag{30}$$

$$[a^0, P] = -iQ \tag{31}$$

$$[a^0, Q] = iP \tag{32}$$

Les opérateurs a^+ et a^- sont connus en physique sous le nom d'opérateurs de création et d'annihilation, et les deux relation équivalentes (25) et (30) sont les "relations de commutation d'Heisenberg".

L'opérateur a^0 quant à lui porte le nom d'opérateur de nombre.

On va maintenant construire toutes les combinaisons linéaires symétriques de ces opérateurs et montrer que ce sont des opérateurs essentiellement auto-adjoints.

PROPOSITION 4 – *Soient* $\tau \in \mathbb{R}$, $z \in \mathbb{C}$, *l'opérateur* $\tau a^0 + za^+ + \bar{z}a^-$ *est essentiellement auto-adjoint.*

Démonstration

On commence par montrer par récurrence sur n que pour tout $k \in \mathbb{N}$

$$|(\tau a^0 + za^+ + \bar{z}a^-)^n e_k| \le (|\tau| + 2|z|)^n \frac{(n+k)!}{k!}$$

Cette relation étant vraie pour $n = 0$, supposons la vraie pour n, et calculons

$$\begin{aligned}
|(\tau a^0 &+ za^+ + \bar{z}a^-)^{n+1} e_k| \\
&= |(\tau a^0 + za^+ + \bar{z}a^-)^n (\tau k e_k + z\sqrt{k+1}\, e_{k+1} + \bar{z}\sqrt{k}\, e_{k-1})| \\
&\le |\tau| k\, |(\tau a^0 + za^+ + \bar{z}a^-)^n (e_k)| \\
&\quad + |z|\sqrt{k+1}|(\tau a^0 + za^+ + \bar{z}a^-)^n (e_{k+1})| \\
&\quad + |z|\sqrt{k}|(\tau a^0 + za^+ + \bar{z}a^-)^n (e_{k-1})| \\
&\le (|\tau| + 2|z|)^n \big(|\tau| k \frac{(n+k)!}{k!} \\
&\quad + |z|\sqrt{k+1} \frac{(n+k+1)!}{(k+1)!} + |z|\sqrt{k} \frac{(n+k-1)!}{(k-1)!} \big) \\
&\le (|\tau| + 2|z|)^{n+1} \frac{(n+k+1)!}{k!}
\end{aligned}$$

ce qui démontre l'inégalité. On en déduit que les vecteur e_k sont analytiques pour $\tau a^0 + za^+ + \bar{z}a^-$ car

$$\begin{aligned}
\sum_{n=0}^{\infty} |(\tau a^0 + za^+ + \bar{z}a^-)^n e_k| \frac{t^n}{n!} &\le \sum_{n=0}^{\infty} (|\tau| + 2|z|)^n \frac{(n+k)!}{k! n!} t^n \\
&= (1 - t(|\tau| + 2|z|))^{-(k+1)} < +\infty \\
&\quad \text{pour } t \ge 0 \text{ assez petit}
\end{aligned}$$

D'après le théorème de Nelson l'opérateur $\tau a^0 + za^+ + \bar{z}a^-$ est essentiellement auto-adjoint.

Dans la suite, par abus de langage on désignera par $\tau a^0 + za^+ + \bar{z}a^-$ sa fermeture.
REMARQUE– Le calcul ci-dessus nous dit en fait un peu plus. Si on a un espace vectoriel dense dans H formé de vecteurs de la forme $\phi = \sum_k \phi_k e_k$ tels qu'il existe un $\beta > 1$ avec $\sum_k |\phi_k| \beta^k < +\infty$, alors les vecteurs de ce sous-espace sont dans le domaine de la fermeture de $\tau a^0 + za^+ + \bar{z}a^-$ et sont analytiques, donc la restriction de $\tau a^0 + za^+ + \bar{z}a^-$ à ce sous-espace est essentiellement auto-adjointe.

3.2 Modèle gaussien des relations de commutation

On va maintenant calculer les lois, dans l'état vide, des opérateurs auto-adjoints que l'on vient de construire, en commençant par le cas $\tau = 0$. La façon dont les opérateurs a^+, a^- ont été obtenus suggère que les lois de P et Q sont des lois normales centrées réduites. C'est ce que nous allons vérifier en calculant leurs transformées de Fourier au moyen des groupes unitaires à un paramètre e^{itP} et e^{itQ}. Ces deux groupes engendrent un groupe d'opérateurs unitaires qui s'expriment à l'aide des opérateurs de Weyl définis ci-dessous. Ces opérateurs sont obtenus au moyen d'une représentation des opérateurs a^+ et a^- sur l'espace L^2 d'une mesure gaussienne, dans laquelle, $Q = a^+ + a^-$ correspond à l'opérateur de multiplication par la fonction x.
Remarquons que les vecteurs e_k de H sont obtenus à partir de la suite de vecteurs $(a^+ + a^-)^k (e_0)$ par le procédé d'orthogonalisation de Gramm-Schmidt. Dans l'espace $L^2(\mathbf{R}, \nu)$, en itérant l'opérateur de multiplication par la fonction x appliqué à la fonction 1 on obtient la suite des monômes $1, x, x^2, \ldots, x^n, \ldots$. Les fonctions obtenues à partir de cette suite par le procédé de Gramm-Schmidt sont les polynômes d'Hermite, dont on rappelle ci-dessous les principales propriétés.

Les polynômes d'Hermite peuvent être définis par la série génératrice

$$e^{tx - \frac{t^2}{2}} = \sum_{n=0}^{+\infty} \frac{t^n}{n!} \mathcal{H}_n(x) \tag{33}$$

La suite des polynômes d'Hermite $(\mathcal{H}_n)_{n \in \mathbb{N}}$ vérifie

$$\int_{-\infty}^{+\infty} \mathcal{H}_n(x) \mathcal{H}_m(x) \frac{e^{-\frac{x^2}{2}}}{\sqrt{2\pi}} \, dx = \delta_{n,m} \, n! \tag{34}$$

On en déduit la suite de polynômes orthonormale dans $L^2(\mathbf{R}, \nu)$

$$h_n(x) = \frac{\mathcal{H}_n(x)}{\sqrt{n!}} \tag{35}$$

A partir de la série génératrice, on obtient les relations de récurrence

$$x \, h_n(x) = \sqrt{n+1} \, h_{n+1}(x) + \sqrt{n} \, h_{n-1}(x) \tag{36}$$

$$\frac{d}{dx} \, h_n(x) = \sqrt{n} \, h_{n-1}(x) \tag{37}$$

Ces relations de récurrence fournissent un modèle pour les opérateurs a^+ et a^-. En effet, soit U l'isomorphisme de H dans $L^2(\mathbf{R}, \nu)$ qui envoie e_k sur h_k alors $a^+ = U^{-1} \circ (x - \frac{d}{dx}) \circ U$ et $a^- = U^{-1} \circ \frac{d}{dx} \circ U$ sur le domaine D. En particulier, l'opérateur Q est conjugué à l'opérateur de multiplication par x, alors que P est conjugué à $\frac{1}{i}(x - 2\frac{d}{dx})$. De même, U transforme l'opérateur $a^0 = a^+ a^-$ en $(-\frac{d^2}{dx^2} + x\frac{d}{dx})$. Cet opérateur est l'opérateur d'Ornstein-Uhlenbeck.

Ce modèle des opérateurs a^+ et a^- sur un espace Gaussien va nous permettre de définir des groupes unitaires à un paramètre.

3.3 Opérateurs de Weyl

DÉFINITION 23 – Soit $z = u + iv \in \mathbf{C}$, on définit un opérateur W_z sur $L^2(\mathbf{R}, \nu)$ par

$$W_z f(x) = f(x - 2u)\, e^{zx - u^2 - iuv}$$

Les opérateurs W_z sont appelés opérateurs de Weyl.

PROPOSITION 5 – Soient $z = u + iv, z' = u' + iv' \in \mathbf{C}$
i) W_z est un opérateur unitaire, $W_0 = Id$.
ii) on a

$$\forall z, z' \in \mathbf{C} \quad W_z W_{z'} = W_{z+z'} e^{i(u'v - uv')} = W_{z+z'} e^{i\Im m(z\bar{z}')} \tag{38}$$

Démonstration
Vérifions ii)

$$
\begin{aligned}
W_z W_{z'} f(x) &= f(x - 2u - 2u') e^{zx - u^2 - iuv} e^{z'(x - 2u) - u'^2 - iu'v'} \\
&= f(x - 2(u + u')) e^{(z+z')x - (u+u')^2 - i(u+u')(v+v') + i(u'v - uv')} \\
&= e^{i(u'v - uv')} W_{z+z'} f(x)
\end{aligned}
$$

Pour i) on vérifie tout d'abord que $W_0 = Id$, puis on calcule

$$
\begin{aligned}
< W_z(f), W_z(g) > &= \int_{-\infty}^{+\infty} f(x - 2u)\bar{g}(x - 2u) e^{zx - u^2 - iuv} e^{\bar{z}x - u^2 + iuv} \frac{e^{-\frac{x^2}{2}}}{\sqrt{2\pi}}\, dx \\
&= \int_{-\infty}^{+\infty} f(x - 2u)\bar{g}(x - 2u) \frac{e^{-\frac{(x-2u)^2}{2}}}{\sqrt{2\pi}}\, dx \\
&= < f, g >
\end{aligned}
$$

ce qui montre que W_z est isométrique, et comme d'après ii) $W_z W_{-z} = W_{-z} W_z = Id$ on voit que W_z est inversible donc unitaire.

COROLLAIRE– Pour tout $z \in \mathbf{C}$, $(W_{tz})_{t \in \mathbf{R}}$ est un groupe d'opérateurs unitaires, fortement continu, dont le générateur est $\frac{1}{i}(za^+ - \bar{z}a^-)$.

Démonstration
La continuité de $t \to W_{tz}f$ se vérifie aisément pour f continue à support compact par le théorème de Lebesgue. Ces fonctions formant un sous-espace dense dans $L^2(\mathbf{R}, \nu)$ on en déduit la continuité forte de $(W_{tz})_{t \in \mathbf{R}}$.

Calculons le générateur infinitésimal de ce groupe unitaire. Pour tout polynôme p on a

$$\frac{W_{tz}p(x) - p(x)}{it} \rightarrow_{t\to 0} \left(-2u\frac{d}{dx}p(x) + zxp(x)\right) = \frac{1}{i}(za^+ - \bar{z}a^-)(p)(x)$$

comme $\frac{1}{i}(za^+ - \bar{z}a^-)$ est essentiellement auto-adjoint sur l'espace des polynômes, il coïncide avec le générateur de $(W_{tz})_{t\in\mathbf{R}}$.

COROLLAIRE– La variable $\frac{1}{i}(za^+ - \bar{z}a^-)$ suit une loi gaussienne centrée de variance $|z|^2$ dans l'état vide.

Démonstration
Calculons la transformée de Fourier de cette loi

$$< e^{it(za^+ + \bar{z}a^-)}\Omega, \Omega > = < W_{tz}h_0, h_0 >$$

$$= \int_{-\infty}^{+\infty} e^{tzx - t^2(u^2 + iuv)} \frac{e^{-\frac{x^2}{2}}}{\sqrt{2\pi}} dx$$

$$= e^{\frac{1}{2}t^2 z^2} e^{-t^2(u^2 + iuv)}$$

$$= e^{-\frac{1}{2}t^2|z|^2}$$

d'où le résultat.

La relation (38) entraine la relation suivante entre les groupes à un paramètre engendrés par les opérateurs P et Q

$$e^{itP}e^{isQ} = e^{-2ist}e^{isQ}e^{itP}$$

(remarquer que $W_t = e^{itP}$ et $W_{is} = e^{isQ}$).
Cette dernière relation permet de retrouver la relation (30), en effet en dérivant on obtient

$$\frac{d}{dt}\frac{d}{ds}_{s=t=0} e^{-itP}e^{isQ}e^{itP} = [P, Q] = 2iId$$

Les relations de commutations (38) entre opérateurs de Weyl sont donc une version "intégrée" des relations de commutation (25) et (30) d'Heisenberg.
Rappelons que le *groupe d'Heisenberg* est $\mathbb{C} \times \mathbf{R}$ muni de la loi de groupe

$$(z, t) \star (z', t') = (z + z', t + t' + \Im m(z\bar{z}'))$$

La relation de commutation de Weyl (38) signifie que $(z, t) \mapsto W_z e^{it}$ est une représentation du groupe d'Heisenberg par des opérateurs unitaires sur $L^2(\mathbf{R}, \nu)$.

3.4 Vecteurs exponentiels

Avant de calculer la loi des opérateurs $\tau a^0 + za^+ + \bar{z}a^-$ pour $\tau \neq 0$ (on peut toujours se ramener à $\tau = 1$), introduisons une définition qui sera très utile quand nous étudierons les espaces de Fock.

DÉFINITION 24 – *Soit $\alpha \in \mathbb{C}$, on pose*

$$\xi(\alpha) = \sum_{k=0}^{\infty} \frac{\alpha^k}{\sqrt{k!}} h_k = e^{\alpha x - \frac{\alpha^2}{2}} = W_\alpha(h_0)e^{\frac{1}{2}|\alpha|^2} \tag{39}$$

Les vecteurs $\xi(\alpha)$ sont appelés vecteurs exponentiels, et on note Ξ le sous-espace vectoriel qu'ils engendrent.

PROPOSITION 6 – *Les vecteurs exponentiels vérifient les propriétés suivantes*
i) $< \xi(\alpha), \xi(\alpha') > = e^{\alpha\bar{\alpha}'}$
ii) Ξ *est dense dans $L^2(\mathbf{R}, \nu)$, et les vecteurs $\xi(\alpha)$ sont linéairement indépendants.*
iii) $W_z(\xi(\alpha)) = \xi(\alpha + z)e^{-\alpha\bar{z} - \frac{1}{2}|z|^2}$

Démonstration
i) $< \xi(\alpha), \xi(\alpha') > = \sum_{n=0}^{\infty} \frac{\alpha^k \bar{\alpha}'^k}{k!} = e^{\alpha\bar{\alpha}'}$
ii) Soient $\alpha_0, \ldots, \alpha_n \in \mathbf{C}$. Les projections orthogonales de $\xi(\alpha_0), \ldots, \xi(\alpha_n)$ sur le sous-espace engendré par $h_0, \ldots, h_n$ ont pour déterminant dans la base $h_0, \ldots, h_n$

$$
\begin{vmatrix}
1 & 1 & \ldots & 1 \\
\alpha_0 & \alpha_1 & \ldots & \alpha_n \\
& & \ldots & \\
\frac{\alpha_0^n}{\sqrt{n!}} & \frac{\alpha_1^n}{\sqrt{n!}} & \ldots & \frac{\alpha_n^n}{\sqrt{n!}}
\end{vmatrix}
$$

Ce déterminant est un multiple d'un déterminant de Vandermonde. Il ne s'annule que si deux des α_k sont égaux, par conséquent les vecteurs $\xi(\alpha_0), \ldots, \xi(\alpha_n)$ forment un système libre si les α_k sont distincts.

La densité de Ξ provient de l'égalité $h_k = \frac{1}{\sqrt{k!}} \frac{d^k}{d\alpha^k} \xi(\alpha)_{|\alpha=0}$ qui montre que les vecteurs h_k sont dans la fermeture de Ξ.
iii) $W_z(e^{\alpha x - \frac{\alpha^2}{2}}) = e^{\alpha(x - 2u) - \frac{\alpha^2}{2}} e^{zx - u^2 - iuv}$
$= e^{(\alpha+z)x - \frac{(\alpha+z)^2}{2}} e^{-\alpha\bar{z} - \frac{1}{2}|z|^2} = \xi(\alpha + z)e^{-\alpha\bar{z} - \frac{1}{2}|z|^2}$

D'après la remarque qui suit la proposition (4), les vecteurs de Ξ sont analytiques pour les opérateurs de la forme $\tau a^0 + za^+ + \bar{z}a^-$, par conséquent, les restrictions de ces opérateurs à Ξ sont essentiellement auto-adjointes.
L'action de a^+ et a^- sur les vecteurs exponentiels prend une forme simple.

$$a^+(\xi(\alpha)) = \frac{d}{d\alpha}\xi(\alpha) \tag{40}$$

$$a^-(\xi(a)) = \alpha\xi(\alpha) \tag{41}$$

3.5 Opérateur de nombre et loi de Poisson

Nous allons maintenant examiner les variables $\tau a^0 + za^+ + \bar{z}a^-$ avec $\tau \neq 0$. L'opérateur de nombre a^0 est donné par sa décomposition spectrale, les vecteurs e_k forment une base de vecteurs propres de a^0. Le groupe unitaire à un paramètre qu'il engendre est donc donné par la formule:

$$e^{ita^0} h_k = e^{itk} h_k$$

L'action sur les vecteurs exponentiels est obtenue facilement, de $a^0 = a^+ a^-$ on tire

$$a^0 \xi(\alpha) = \alpha\frac{d}{d\alpha}\xi(\alpha) \tag{42}$$

et

$$e^{ita^0}\xi(\alpha) = \xi(e^{it}\alpha) \tag{43}$$

On va en déduire les relations de commutation entre ce groupe unitaire et les opérateurs de Weyl.

PROPOSITION 7 –

$$e^{ita^0}W_z e^{-ita^0} = W_{e^{it}z} \tag{44}$$

$$W_{-z}e^{ita^0}W_z = e^{it(a^0+za^+ +\bar{z}a^- +|z|^2)} \tag{45}$$

Démonstration

Pour la première formule, considérons un vecteur exponentiel $\xi(\alpha)$, et calculons

$$\begin{aligned}
e^{ita^0}W_z e^{-ita^0}\big(\xi(\alpha)\big) &= e^{ita^0}W_z\big(\xi(e^{-it}\alpha)\big)\\
&= e^{ita^0}\Big(\xi(e^{-it}\alpha + z)e^{-\bar{z}\alpha e^{-it}-\frac{|z|^2}{2}}\Big)\\
&= \xi(\alpha + ze^{it})e^{-\bar{z}\alpha e^{-it}-\frac{|z|^2}{2}}\\
&= W_{e^{it}z}\xi(\alpha)
\end{aligned}$$

les vecteurs exponentiels forment une partie totale, donc les deux opérateurs unitaires coïncident ce qui montre la première partie.

La famille $(W_{-z}e^{ita^0}W_z)_{t\in\mathbb{R}}$ est un groupe à un paramètre d'opérateurs unitaires. Calculons son générateur.

$$\begin{aligned}
\frac{1}{i}\frac{d}{dt}_{t=0}(W_{-z}e^{ita^0}W_z(\xi(\alpha)) &= \frac{1}{i}\frac{d}{dt}_{t=0}\big(\xi(e^{it}(\alpha + z) - z)e^{-\alpha\bar{z}-\frac{1}{2}|z|^2}e^{e^{it}(\alpha+z)\bar{z}-\frac{1}{2}|z|^2}\big)\\
&= (\alpha + z)\frac{d}{d\alpha}(\xi(\alpha)) + (\bar{z}\alpha + |z|^2)\xi(\alpha)\\
&= (a^0 + za^+ + \bar{z}a^- + |z|^2)(\xi(\alpha))
\end{aligned}$$

Le générateur de $(W_{-z}e^{ita^0}W_z)_{t\in\mathbb{R}}$ coincide avec $a^0+za^+ +\bar{z}a^- +|z|^2$ sur le domaine Ξ, or $a^0 + za^+ + \bar{z}a^- + |z|^2$ est essentiellement auto-adjoint sur ce domaine, d'après la remarque qui suit la proposition (4) donc ces deux opérateurs sont égaux, et en appliquant le théorème de Stone on a la seconde formule.

La proposition précédente nous permet d'obtenir la loi de la variable $a^0 + za^+ + \bar{z}a^- + |z|^2$ dans l'état vide:

PROPOSITION 8 – *La variable* $a^0+za^+ +\bar{z}a^- +|z|^2$ *suit une loi de Poisson de paramètre* $|z|^2$.

Démonstration

je vais donner deux façons de calculer cette loi, afin d'illustrer les principes de base des probabilités non-commutatives. La première consiste à calculer la transformée de Fourier

$$< e^{it(a^0+za^+ +\bar{z}a^- +|z|^2)}\Omega, \Omega > = < W_{-z}e^{ita^0}W_z h_0, h_0 >$$
$$= < e^{ita^0}W_z h_0, W_z h_0 >$$
$$= e^{-|z|^2} < e^{ita^0}\xi(z), \xi(z) >$$
$$= e^{-|z|^2} < \xi(e^{it}z), \xi(z) >$$
$$= e^{|z|^2(e^{it}-1)}$$

Pour la seconde, on remarque que $a^0 + za^+ + \bar{z}a^- + |z|^2 = W_{-z}a^0 W_z$, et on applique le lemme suivant.

LEMME – Soient A un opérateur auto-adjoint sur H, U un opérateur unitaire et $x \in H$ de norme 1, la loi de $U^{-1}AU$ dans l'état pur x est égale à la loi de A dans l'état pur Ux.

La loi de a^0 dans l'état $W_z(h_0)$ est facile à calculer, le spectre de a^0 est $\mathbb{N}$ et le vecteur h_k est un vecteur propre de a^0 de valeur propre k, donc la loi est portée par les entiers positifs et donne comme probabilité à k

$$| < W_z(h_0), h_k > |^2 = \frac{|z|^{2k}}{k!}e^{-|z|^2}$$

on reconnait la loi de Poisson de paramètre $|z|^2$.

Les relations de commutation de Weyl fournissent une représentation du groupe d'Heisenberg. Si on tient compte de la relation (44) on obtient une représentation d'un groupe qui contient le groupe d'Heisenberg. Ce groupe est $U(1) \times \mathbb{C} \times \mathbb{R}$ avec la loi de groupe

$$(e^{i\theta}, z, t) \star (e^{i\theta'}, z', t') = (e^{i(\theta+\theta')}, z + e^{i\theta}z', t + t' + \Im m (z\bar{z}'))$$

La relation (44) montre que l'application

$$(e^{i\theta}, z, t) \mapsto W_z e^{i\theta a^0} e^{it}$$

définit une représentation unitaire de ce groupe.
Remarquons également que la relation (44) entraine

$$e^{i\frac{\pi}{2}a^0}e^{itP}e^{-i\frac{\pi}{2}a^0} = e^{itQ}$$

autrement dit les variables P et Q sont unitairement équivalente par la transformation $e^{i\frac{\pi}{2}a^0}$. Cette transformation unitaire de $L^2(\mathbb{R}, \nu)$ n'est autre que la transformation de Fourier.

3.6 Modèle poissonien des relations de commutation

Au cours de ce chapitre, j'ai utilisé l'espace L^2 de la mesure de Gauss pour donner une représentation des relations de commutation d'Heisenberg. On aurait pu également utiliser l'espace L^2 de la mesure de Poisson bien que cela soit moins habituel. Je vais indiquer brièvement dans ce paragraphe comment s'interprètent alors les opérateurs de création, d'annihilation et de nombre.

On considère la mesure de Poisson de paramètre 1, sur $\mathbb{N}$, donnée par la formule $\mu(\{k\}) = \frac{e^{-1}}{k!}$.

Les polynômes orthogonaux par rapport à cette mesure sont les polynômes de Charlier (voir Chihara [16]), $(C_n)_{n \in \mathbb{N}}$ définis par la série génératrice

$$\sum_{n=0}^{\infty} C_n(x) \frac{w^n}{n!} = e^{-x}(1 + w)^x \tag{46}$$

Ils vérifient les relations de récurrence

$$C_{n+1}(x) = (x - n)C_n(x) - nC_{n-1}(x) \tag{47}$$

$$C_n(x + 1) - C_n(x) = nC_{n-1}(x) \tag{48}$$

Les polynômes $c_n = \sqrt{n!}\, C_n$ forment une base orthonormale de $L^2(\mathbb{N}, \mu)$. Introduisons l'opérateur de différence finie Δ défini par

$$\Delta f(x) = f(x + 1) - f(x)$$

Il possède un adjoint Δ^* donné par $\Delta^* f(x) = xf(x - 1) - f(x)$. Les relations de récurrence des polynomes de Charlier entrainent alors que

$$\Delta^* c_n = \sqrt{n + 1}\, c_{n+1} \tag{49}$$

$$\Delta c_n = \sqrt{n}\, c_{n-1} \tag{50}$$

autrement dit, les opérateurs Δ^* et Δ donnent un modèle des opérateurs de création et d'annihilation. L'opérateur de multiplication par la fonction x est donné par la formule $(\Delta^* + Id)(\Delta + Id)$ et par construction, il suit une loi de Poisson de paramètre 1 dans l'état vide.

3.7 Théorème de Stone-von Neumann

La fin du chapitre 3 est constituée de compléments qui ne sont pas indispensables pour la suite du cours.

Les opérateurs de Weyl vérifient les relations de commutation (38), qui sont une version intégrée des relations d'Heisenberg, et définissent ainsi une représentation du groupe d'Heisenberg. Nous allons voir que cette représentation est irréductible, et que toute autre représentation en est un multiple, c'est le contenu du théorème de Stone-von Neumann démontré dans ce paragraphe. Ensuite, on exploitera cette propriété des relations de commutation pour introduire de façon naturelle de nouvelles variables aléatoires et faire quelques calculs de lois. Ces variables jouent un rôle important en optique dans la théorie des états dits "comprimés" de la lumière (cf S.Reynaud [71]).

Voici l'énoncé du théorème de Stone-von Neumann.

THÉORÈME 6 – Soit $(\tilde{W}_z)_{z \in \mathbb{C}}$ une famille fortement continue d'opérateurs unitaires sur un espace de Hilbert $\tilde{H}$ telle que

$$\forall z, z' \in \mathbb{C} \quad \tilde{W}_z \tilde{W}'_z = \tilde{W}_{z+z'} e^{i\Im m(z\bar{z}')} \tag{51}$$

alors il existe un espace de Hilbert $\tilde{H}_0$ et isomorphisme d'espaces de Hilbert

$$U : \tilde{H} \to L^2(\mathbf{R}, \nu) \otimes \tilde{H}_0$$

tel que $\tilde{W}_z = U^*(W_z \otimes Id)U$.
D'autre part, les opérateurs W_z opèrent de façon irréductible sur $L^2(\mathbf{R}, \nu)$, c'est à dire que les seuls sous-espaces fermés invariants par tous ces opérateurs sont $\{0\}$ et $L^2(\mathbf{R}, \nu)$.

Démonstration
Il existe de nombreuses démonstrations du théorème de Stone-von Neumann (cf [73], [48], [70]). En voici une de nature algébrique.
Commençons par remarquer que la relation (51) entraine que $\tilde{W}_0 = Id$ et pour tout $z \in \mathbf{C}$, $\tilde{W}_z^* = \tilde{W}_{-z}$.
On va établir les deux lemmes suivants:

LEMME 1 – L'opérateur $\tilde{P} = \frac{1}{2\pi} \int_{\mathbf{C}} \tilde{W}_z e^{-\frac{1}{2}|z|^2} dz$ est auto-adjoint, et pour tout $\zeta \in \mathbf{C}$ on a

$$\tilde{P}\tilde{W}_\zeta \tilde{P} = e^{-\frac{1}{2}|\zeta|^2} \tilde{P}$$

En particulier (en prenant $\zeta = 0$), $\tilde{P}$ est un projecteur.

Démonstration
Cet opérateur est bien défini et borné car $z \to \tilde{W}_z$ est fortement continue bornée. Il est auto-adjoint car $\tilde{P}^* = \frac{1}{2\pi} \int_{\mathbf{C}} \tilde{W}_{-z} e^{-\frac{1}{2}|z|^2} dz = \frac{1}{2\pi} \int_{\mathbf{C}} \tilde{W}_z e^{-\frac{1}{2}|z|^2} dz = \tilde{P}$. Calculons

$$\tilde{P}\tilde{W}_\zeta \tilde{P} = (\frac{1}{2\pi})^2 \int\int_{\mathbf{C}^2} \tilde{W}_z \tilde{W}_\zeta \tilde{W}_{z'} e^{-\frac{1}{2}(|z|^2 + |z'|^2)} dz \, dz'$$

$$= (\frac{1}{2\pi})^2 \int\int_{\mathbf{C}^2} \tilde{W}_{z+\zeta+z'} e^{-\frac{1}{2}(|z|^2 + |z'|^2) + \bar{z}\zeta - z\bar{\zeta} + \bar{z}z' - z\bar{z}' + \bar{\zeta}z' - \zeta\bar{z}'} dz \, dz'$$

$$= (\frac{1}{2\pi})^2 \int\int_{\mathbf{C}^2} \tilde{W}_{z+\zeta+z'} e^{-\frac{1}{2}|z+\zeta+z'|^2 + (\bar{z}+\bar{\zeta})(z'+\zeta) - \frac{1}{2}|\zeta|^2} dz \, dz'$$

$$= (\frac{1}{2\pi})^2 \int\int_{\mathbf{C}^2} \tilde{W}_{z''} e^{-\frac{1}{2}|z''|^2 + (\bar{z}'' - \bar{\zeta} - \bar{z}''')z''' - \frac{1}{2}|\zeta|^2} dz'' \, dz'''$$

en faisant le changement de variables: $z'' = z + \zeta + z'$, $z''' = z' + \zeta$

$$= (\frac{1}{2\pi}) \int_{\mathbf{C}} \tilde{W}_{z''} e^{-\frac{1}{2}|z''|^2 - \frac{1}{2}|\zeta|^2} dz''$$

en intégrant par rapport à z''' et en utilisant la formule $\frac{1}{2\pi} \int_{\mathbf{C}} e^{tz - \bar{z}z} dz = 1$

$$= e^{-\frac{1}{2}|\zeta|^2} \tilde{P}$$

ce qui termine la démonstration du lemme.

LEMME 2 – $P = \frac{1}{2\pi} \int_{\mathbf{C}} W_z e^{-\frac{1}{2}|z|^2} dz$ est le projecteur orthogonal sur la droite engendrée par h_0 dans $L^2(\mathbf{R}, \nu)$.

Démonstration
Soit $f \in L^2(\mathbf{R}, \nu)$, on a

$$Pf(x) = \frac{1}{2\pi} \int_{-\infty}^{+\infty} \int_{-\infty}^{+\infty} f(x-2u)e^{x(u+iv)-u(u+iv)-\frac{1}{2}(u^2+v^2)} \, du \, dv$$

$$= \frac{1}{\sqrt{2\pi}} \int_{-\infty}^{+\infty} f(x-2u)e^{-\frac{1}{2}(x-u)^2+xu-\frac{3}{2}u^2} \, du$$

$$= \frac{1}{\sqrt{2\pi}} \int_{-\infty}^{+\infty} f(x-2u)e^{-\frac{1}{2}(x-2u)^2} \, du$$

$$= < f, h_0 > h_0(x)$$

COROLLAIRE – Les opérateurs W_z opèrent de façon irréductible.

Démonstration

Soit K un sous-espace fermé invariant par les W_z. Le projecteur orthogonal sur K commute avec les opérateurs W_z, donc également avec les opérateurs $W_z P W_{-z}$ pour $z \in \mathbb{C}$. Or, d'après le lemme 2, $W_z P W_{-z}$ est le projecteur orthogonal sur la droite engendrée par le vecteur $W_z(h_0)$. Ceci entraine que pour tout $z \in \mathbb{C}$ on a soit $W_z(h_0) \in K$ soit $W_z(h_0) \in K^\perp$. Comme $< W_z(h_0), W_{z'}(h_0) > \neq 0$ on voit que soit K soit son orthogonal contient tous les vecteurs $W_z(h_0)$ donc tous les vecteurs $\xi(z)$. Ces vecteurs forment une partie totale de $L^2(\mathbb{R}, \nu)$, et K étant fermé on a $K = H$ ou $K = \{0\}$.

LEMME 3 – Pour tous $z_1, z_2 \in \mathbb{C}$, $x_1, x_2 \in \tilde{H}$ on a

$$< \tilde{W}_{z_1} \tilde{P} x_1, \tilde{W}_{z_2} \tilde{P} x_2 > = < \tilde{P} x_1, \tilde{P} x_2 > < W_{z_1} h_0, W_{z_2} h_0 >$$

Démonstration

$$< \tilde{W}_{z_1} \tilde{P} x_1, \tilde{W}_{z_2} \tilde{P} x_2 > = < \tilde{P} \tilde{W}_{-z_2} \tilde{W}_{z_1} \tilde{P} x_1, x_2 >$$

$$= e^{-i\Im m(z_2 \bar{z}_1)} < \tilde{P} \tilde{W}_{z_1-z_2} \tilde{P} x_1, x_2 >$$

$$= e^{-i\Im m(z_2 \bar{z}_1)-\frac{1}{2}|z_1-z_2|^2} < \tilde{P} x_1, x_2 > \quad \text{d'après le lemme 1}$$

$$= < \tilde{P} x_1, \tilde{P} x_2 > < W_{z_1} h_0, W_{z_2} h_0 >$$

Le lemme 3 entraine que l'application

$$U : \Xi \otimes Im(\tilde{P}) \to \tilde{H}$$

$$W_z h_0 \otimes x \mapsto \tilde{W}_z x$$

se prolonge en une isométrie de $L^2(\mathbb{R}, \nu) \otimes Im(\tilde{P})$ dans $\tilde{H}$.

Montrons que son image est $\tilde{H}$ tout entier. Supposons que $y \in \tilde{H}$ est orthogonal à $\tilde{W}_z(Px) \; \forall x \in \tilde{H} \; z \in \mathbb{C}$. Pour tout $\zeta \in \mathbb{C}$ on a:

$$< \tilde{W}_\zeta \tilde{P} \tilde{W}_{-\zeta} x, y > = 0$$

$$= \int_{\mathbb{C}} < W_z x, y > e^{-\frac{1}{2}|z|^2+2i\Im m(z\bar{\zeta})} dz$$

La fonction $z \mapsto < W_z x, y >$ est continue, donc elle est nulle (injectivité de la transformée de Fourier) et $< x, y > = 0 \; \forall x \in \tilde{H}$, d'où $y = 0$.

L'application U est bien l'isomorphisme cherché, on a $\tilde{W}_z = U^*(W_z \otimes Id)U$ par construction, ce qui termine la démonstration.

3.8 Conséquences du théorème de Stone-von Neumann

Supposons, avec les notations du théorème 6, que les opérateurs $\tilde{W}_z$ agissent de façon irréductible, alors l'espace $\tilde{H}_0$ est de dimension 1, et U est un isomorphisme entre $\tilde{H}$ et $L^2(\mathbf{R}, \nu)$.

Soit $\tau : \mathbf{C} \to \mathbf{C}$ une application $\mathbf{R}$-linéaire, alors $\Im m\,(\tau(z)\tau(\bar{z}')) = \det(\tau)\Im m\,(z\bar{z}')$ donc l'application τ vérifie

$$\forall z, z' \in \mathbf{C} \; \Im m\,(\tau(z)\tau(\bar{z}')) = \Im m\,(z\bar{z}')$$

si et seulement si $\det(\tau) = 1$, et les opérateurs $\tilde{W}_z = W_{\tau(z)}$ vérifient alors les hypothèses du théorème de Stone-von Neumann. Ils agissent de façon irréductible sur $L^2(\mathbf{R}, \nu)$ ce qui entraine l'existence d'un opérateur unitaire U_τ tel que $W_{\tau(z)} = U_\tau^* W_z U_\tau$ pour tout $z \in \mathbf{C}$. Si V_τ est un autre opérateur unitaire vérifiant cette égalité, alors $U_\tau V_\tau^*$ commute avec W_z pour tout $z \in \mathbf{C}$ et donc c'est un multiple de l'identité, par irréductibilité de la représentation. L'opérateur U_τ est donc déterminé à un nombre complexe de module 1 près.

Je vais donner ci-dessous deux exemples de groupes à un paramètre de telles transformations.

Le premier exemple est celui des applications $\tau_t(z) = ze^{it}$. D'après les relations (44) on a

$$e^{ita^0} W_z e^{-ita^0} = W_{ze^{it}}$$

ce qui montre que l'opérateur unitaire U_{τ_t} correspondant est e^{ita^0}.

Le second exemple est donné par les applications $\tau_t(u+iv) = e^t u + ie^{-t} v$. L'opérateur U_{τ_t} donné par la formule suivante

$$U_{\tau_t} f(x) = f(e^t x) e^{\frac{t}{2} - \frac{1}{4}(e^{2t}-1)x^2}$$

vérifie pour $z = u + iv$

$$
\begin{aligned}
U_{\tau_t}^* W_{u+iv} U_{\tau_t} f(x) &= (W_{u+iv} U_{\tau_t} f)(e^{-t} x) e^{-\frac{t}{2} - \frac{1}{4}(e^{-2t}-1)x^2)} \\
&= U_{\tau_t} f(e^{-t}(x-2u)) e^{ze^{-t}x - u^2 - iuv} e^{-\frac{t}{2} - \frac{1}{4}(e^{-2t}-1)x^2)} \\
&= f(x - 2e^t u) e^{ze^{-t}x - u^2 - iuv} e^{-\frac{t}{2} - \frac{1}{4}(e^{-2t}-1)x^2} e^{\frac{t}{2} - \frac{1}{4}(e^{2t}-1)(e^{-t}x-2u)^2} \\
&= f(x - 2e^t u) e^{(e^t u + ie^{-t} v)x - e^{2t}u^2 - iuv} \\
&= W_{e^t u + ie^{-t} v} f(x)
\end{aligned}
$$

Les opérateurs U_{τ_t} forment un groupe à un paramètre, dont nous allons calculer le générateur.

Pour un polynome p en la variable x on a

$$\lim_{t \to 0} \frac{U_{\tau_t} p(x) - p(x)}{it} = \frac{1}{i}(x \frac{d}{dx} p(x) - (\frac{x^2}{2} + 1)p(x))$$

On en déduit que, sur le domaine des polynômes, ce générateur coïncide avec $\frac{1}{2}(PQ + QP)$.

Comme dans la proposition 4 on montre facilement que $\frac{1}{2}(PQ+QP)$ est essentiellement auto-adjoint sur le domaine formé des polynômes, et sa fermeture auto-adjointe
a donc une loi dans l'état vide dont la transformée de Fourier est

$$< e^{\frac{it}{2}(PQ+QP)}\Omega, \Omega > \; = \; < U_{\tau_t}\Omega, \Omega >$$

$$= \int_{-\infty}^{+\infty} e^{\frac{t}{2}-\frac{1}{4}(e^{2t}-1)x^2} \frac{e^{-\frac{1}{2}x^2}}{\sqrt{2\pi}}\,dx$$

$$= \frac{1}{\sqrt{\mathrm{ch}t}}$$

Terminons ce paragraphe par le calcul de la loi de $\alpha P^2 + \beta Q^2$ $(\alpha, \beta \in \mathbf{R}_+^*)$. Les
opérateurs $P^2 + Q^2$ et $PQ + QP$ vérifient la relation

$$[P^2 + Q^2, PQ + QP] = 8i(P^2 - Q^2)$$

on en déduit que

$$e^{-i\frac{t}{2}(PQ+QP)}(P^2 + Q^2)e^{i\frac{t}{2}(PQ+QP)} = e^{-4t}P^2 + e^{4t}Q^2$$

(pour vérifier cette égalité, dériver deux fois par rapport à t et utiliser la relation
de commutation précédente). On a donc $\alpha P^2 + \beta Q^2 = \sqrt{\alpha\beta}U_{\tau_{-t}}(P^2 + Q^2)U_{\tau_t}$ avec
$\beta = \alpha e^{8t}$. Le calcul de la loi de $\alpha P^2 + \beta Q^2$ dans l'état vide est ramené à celui de loi
de $P^2 + Q^2$ dans l'état $U_{\tau_t}\Omega$, or $P^2 + Q^2 = 4a^0 + 2Id$ donc il suffit de calculer la
loi de a^0 dans l'état $U_{\tau_t}\Omega$. L'opérateur a^0 est donné par sa décomposition spectrale,
le vecteur h_n étant vecteur propre de valeur propre n. La loi de a^0 dans l'état pur
$U_{\tau_t}\Omega$ est donc la loi sur $\mathbb{N}$ donnant la masse $|<h_n, U_{\tau_t}\Omega>|^2$ au point n. On a

$$\sum_0^\infty \frac{u^n}{\sqrt{n!}} <h_n, U_{\tau_t}\Omega> = (2\pi)^{-\frac{1}{2}}\int_{-\infty}^{+\infty} e^{ux-\frac{u^2}{2}-\frac{x^2}{4}}e^{\frac{t}{2}-(e^{2t}-1)\frac{x^2}{4}}e^{-\frac{x^2}{2}}\,dx$$

$$= \frac{1}{\sqrt{\mathrm{ch}t}}e^{\frac{u^2}{2}\mathrm{th}t}$$

d'où l'on tire la valeur de la probabilité de n qui est 0 si n est impair, ct $\frac{1}{\mathrm{ch}t}(\frac{\mathrm{th}t}{2})^n\frac{n!}{(\frac{n}{2}!)^2}$
sinon.

Chapitre 4

Espaces de Fock

4.1 Puissances symétriques d'un espace de Hilbert

Nous avons construit dans le chapitre précédent des variables de Gauss et de Poisson à l'aide d'opérateurs de création, annihilation et nombre. Afin de pouvoir étudier non plus seulement des variables mais des processus, nous allons introduire la notion d'espace de Fock qui est un moyen naturel de considérer un produit de copies indépendantes de l'espace du chapitre 3. Nous allons retrouver dans ce contexte des vecteurs exponentiels, opérateurs d'annihilation, de création, de Weyl etc ....
On commence par des préliminaires algébriques au paragraphes 4.1 et 4.2, puis on aborde les probabilités avec les paragraphes 4.3 et suivants.

Soit K un espace de Hilbert, et $n \in \mathbb{N}^*$, le groupe $\mathfrak{S}_n$ des permutations de $[1, \ldots, n]$ agit sur $K^{\otimes n}$ par

$$\sigma\big(x_1 \otimes \ldots \otimes x_n\big) = x_{\sigma(1)} \otimes \ldots \otimes x_{\sigma(n)}$$

DÉFINITION 25 – *On note $K^{\circ n}$ le sous-espace de $K^{\otimes n}$ des éléments invariants par $\mathfrak{S}_n$. C'est un sous-espace de Hilbert de $K^{\otimes n}$ appelé puissance symétrique n^e de K et l'opérateur $\mathsf{s} = \frac{1}{n!} \sum_{\sigma \in \mathfrak{S}_n} \sigma$ est le projecteur orthogonal sur ce sous-espace.*

Soient $h_1, \ldots, h_n$, $k_1, \ldots, k_n \in K$ on pose

$$k_1 \circ \ldots \circ k_n = \sqrt{n!}\, \mathsf{s}(k_1 \otimes \ldots \otimes k_n) \tag{52}$$

On a

$$< h_1 \circ \ldots \circ h_n, k_1 \circ \ldots \circ k_n > = \sum_{\sigma \in \mathfrak{S}_n} \prod_{i=1}^{n} < h_i, k_{\sigma(i)} > \tag{53}$$

L'espace vectoriel engendré par les vecteurs $k^{\circ n}$ est $K^{\circ n}$.

DÉFINITION 26 – *Soit K un espace de Hilbert on appelle espace de Fock construit sur K l'espace de Hilbert*

$$\Gamma(K) = \bigoplus_{n=0}^{\infty} K^{\circ n}$$

où la somme directe est une somme Hilbertienne, et par convention $K^{\circ 0}$ est un espace de Hilbert de dimension 1, engendré par un vecteur unitaire noté Ω et appelé "vecteur vide".

On notera $\Gamma_0(K)$ la somme directe non complétée $\bigoplus_{n=0}^{\infty} K^{\circ n}$.

DÉFINITION 27 – *Soit $k \in K$ on pose $k^{\circ 0} = \Omega$ et*

$$\xi(k) = \sum_{n=0}^{\infty} \frac{k^{\circ n}}{n!}$$

Les vecteurs de la forme $\xi(k)$ pour $k \in K$ sont appelés vecteurs exponentiels, et on note $\Xi(K)$ l'espace vectoriel qu'ils engendrent.

Si $K = \mathbb{C}.u$ est de dimension 1 avec $|u| = 1$, alors en identifiant $u^{\circ n}$ avec $\sqrt{n!}e_n$ on obtient une isométrie entre $\Gamma(K)$ et l'espace de Hilbert H du chapitre 3. Les vecteurs exponentiels $\xi(\alpha)$ du chapitre 3 correspondent aux vecteurs $\xi(\alpha.u)$ ci-dessus.

PROPOSITION 9 – *Les vecteurs exponentiels forment un système libre et total dans $\Gamma(K)$, et on a*

$$\forall h, k \in K \quad < \xi(h), \xi(k) > = e^{<h,k>} \tag{54}$$

Démonstration
La formule résulte de la définition des vecteurs exponentiels. Pour tout $k \in K$ on a $\frac{d^n}{d\varepsilon^n}\xi(\varepsilon k)_{\varepsilon=0} = k^{\circ n}$, donc les vecteurs $k^{\circ n}$ sont dans l'adhérence de $\Xi(K)$, et ces vecteurs engendrent Γ_0 tout entier.
La liberté du système $\xi(k)$ se démontre de manière analogue à celle des vecteurs exponentiels du chapitre 3.

On déduit de la proposition 9 que ξ est une application continue, en effet, on a

$$|\xi(h) - \xi(k)|^2 = e^{<h,h>} - e^{<h,k>} - e^{<k,h>} + e^{<k,k>}$$

qui tend vers 0 quand $h \to k$. En particulier, si $S \subset K$ est une partie dense, l'espace $\Xi(S)$ engendré par les vecteurs exponentiels d'éléments de S est dense dans $\Gamma(K)$.

PROPOSITION 10 – *Soit $K = K_1 \oplus K_2$ une décomposition orthogonale, l'application*

$$\Xi(K) \to \Xi(K_1) \otimes \Xi(K_2)$$
$$\xi(k) \mapsto \xi(k_1) \otimes \xi(k_2)$$

(où $k = k_1 + k_2$ est la décomposition de k suivant $K_1 \oplus K_2$) se prolonge en une isométrie de $\Gamma(K)$ sur $\Gamma(K_1) \otimes \Gamma(K_2)$.

Démonstration
D'après la formule (54) on a

$$< \xi(k), \xi(l) > = e^{<k,l>} = e^{<k_1,l_1>+<k_2,l_2>}$$
$$= < \xi(k_1), \xi(l_1) >< \xi(k_2), \xi(l_2) > = < \xi(k_1) \otimes \xi(k_2), \xi(l_1) \otimes \xi(l_2) >$$

qui montre que l'application est une isométrie surjective. Comme $\Xi(K)$ est dense dans $\Gamma(K)$, on conclut.

En particulier, si $K_1 = \mathbb{C}k$ et $K_2 = (\mathbb{C}k)^\perp$ alors $\Gamma(K) \sim H \otimes \Gamma(K_2)$.
La proposition (10) s'étend immédiatement au cas d'une décomposition finie

$$\Gamma(\bigoplus_{i=1}^{n} K_i) \sim \bigotimes_{i=1}^{n} \Gamma(K_i)$$

Si on a une décomposition orthogonale infinie, $K = \bigoplus_{i=1}^{\infty} K_i$ l'isomorphisme

$$\Gamma(\bigoplus_{i=1}^{\infty} K_i) \sim \bigotimes_{i=1}^{\infty} \Gamma(K_i)$$

reste valable, le produit infini $\bigotimes_{i=1}^{\infty} \Gamma(K_i)$, étant défini de la façon suivante.
On considère l'espace vectoriel engendré par les vecteurs de la forme
$u_1 \otimes \ldots \otimes u_n \otimes \ldots$ où $u_n \in \Gamma(K_n)$ est égal à Ω_n (le vecteur vide de $\Gamma(K_n)$) pour
tous les n en dehors d'un nombre fini. C'est un espace préhilbertien pour le produit

$$< h_1 \otimes \ldots \otimes h_n \otimes \ldots, k_1 \otimes \ldots \otimes k_n \otimes \ldots >=< h_1, k_1 > \ldots < h_n, k_n > \ldots$$

(le produit converge car les termes valent 1 à partir d'un certain rang) et sa
complétion est l'espace de Hilbert $\bigotimes_{i=1}^{\infty} \Gamma(K_i)$.
Si (k_i) est une base orthonormale de K, on a l'isomorphisme $\Gamma(K) = \bigotimes_i \Gamma(\mathbb{C}k_i)$.

4.2 Opérateurs de création et d'annihilation

DÉFINITION 28 – Soit $k \in K$ on définit deux opérateurs de domaine $\Gamma_0(K)$ par:

$$a_k^+(k_1 \circ \ldots \circ k_n) = k \circ k_1 \circ \ldots \circ k_n \tag{55}$$

$$a_k^-(k_1 \circ \ldots \circ k_n) = \sum_{i=1}^{n} < k_i, k > k_1 \circ \ldots \circ \widehat{k_i} \circ \ldots \circ k_n \tag{56}$$

Les opérateurs a_k^+ sont appelés opérateurs de création, les a_k^- opérateurs d'annihilation.

a_k^+ dépend linéairement de k alors que a_k^- en dépend antilinéairement.
Ils vérifient la relation d'adjonction

$$\forall u, v \in \Gamma_0(K) \quad < a_k^+(u), v >=< u, a_k^-(v) > \tag{57}$$

en particulier, ils sont fermables et $a_k^+ \subset (a_k^-)^*$, $a_k^- \subset (a_k^+)^*$.
Le domaine $\Gamma_0(K)$ est invariant par a_k^+ et a_k^-.
On vérifie aisément que les vecteurs exponentiels sont dans le domaine de la
fermeture des opérateurs de création et d'annihilation et on a

$$a_k^+(\xi(h)) = \frac{d}{dt}\xi(h + tk)_{t=0} \tag{58}$$

$$a_k^-(\xi(h)) =< h, k > \xi(h) \tag{59}$$

Lorsque $K = \mathbb{C}.u$ est de dimension 1, en utilisant l'identification ci-dessus de $\Gamma(K)$
avec H, on voit que l'opérateur a_{zu}^+ correspond à za^+, et a_{zu}^- à $\bar{z}a^-$. Si $k \in K$, et
$|k| = 1$ dans la décomposition $\Gamma(K) \sim \Gamma(\mathbb{C}k) \otimes \Gamma((\mathbb{C}k)^\perp)$, on a $a_k^\pm \sim a^\pm \otimes Id$. On en
déduit que l'opérateur $a_k^+ + a_k^-$ est essentiellement auto-adjoint sur $\Gamma_0(K)$ et qu'il
suit une loi normale centrée réduite dans l'état Ω. En particulier, si on dispose d'une
base orthonormée $(k_i)_{i \in I}$ de K, les opérateurs $a_{k_i}^+$, $a_{k_i}^-$, $i \in I$ forment une famille de
"copies indépendantes" (dans un sens analogue à celui du chapitre 1) des opérateurs
a^+, a^-.

Les relations de commutation (25) deviennent

$$[a_h^+, a_k^-] = - < h, k > Id \tag{60}$$

Si nous utilisons les opérateurs $Q_u = a_u^+ + a_u^-$ et $P_u = \frac{1}{i}(a_u^+ - a_u^-) = Q_{\frac{1}{i}u}$ ces
relations prennent la forme

$$[Q_u, Q_v] = -2i\Im m < u, v > Id$$

$$[P_u, P_v] = -2i\Im m < u, v > Id \qquad (61)$$

Terminons ce paragraphe en introduisant deux opérateurs que nous utiliserons dans le chapitre suivant.

DÉFINITION 29 – L'opérateur gradient sur l'espace de Fock est la fermeture de l'opérateur de $\Gamma_0(K)$ dans $K \otimes \Gamma(K)$ défini par

$$\forall k \in K \; \forall X, Y \in \Gamma_0(K) \quad < \nabla(X), k \otimes Y > = < a_k^-(X), Y > \qquad (62)$$

La divergence δ est l'adjoint de ∇, c'est donc un opérateur fermé de $Dom(\delta) \subset K \otimes \Gamma(K)$ dans $\Gamma(K)$, et on a

$$\delta(k \otimes X) = a_k^+(X)$$

4.3 Seconde quantification

Soit A un opérateur sur K, l'opérateur $A^{\circ n}$ est défini sur le sous-espace $D(A)^{\circ n}$ de $K^{\circ n}$ engendré par les vecteurs $k_1 \circ \ldots \circ k_n$, $k_i \in D(A)$ par

$$A^{\circ n} k_1 \circ \ldots \circ k_n = Ak_1 \circ \ldots \circ Ak_n$$

Si A est borné, alors $A^{\circ n}$ est borné de norme $\|A\|^n$.

DÉFINITION 30 – La seconde quantification de A, notée $\Gamma(A)$ est l'opérateur de domaine $\bigoplus_{n \in \mathbb{N}} D(A)^{\circ n}$ (somme algébrique) défini par $\Gamma(A) = \bigoplus_{n \in \mathbb{N}} A^{\circ n}$.

Si A est borné de norme ≤ 1, alors $\Gamma(A)$ a une fermeture bornée, de norme ≤ 1, mais si $\|A\| > 1$, $\Gamma(A)$ n'est pas borné.

Si U est unitaire sur K alors $\Gamma(U)$ est unitaire, et si $(U_t)_{t \in \mathbb{R}}$ est un groupe à un paramètre de tels opérateurs, fortement continu, alors $(\Gamma(U_t))_{t \in \mathbb{R}}$ est un groupe unitaire fortement continu sur $\Gamma(K)$. Le générateur de ce groupe s'exprime à l'aide du générateur V de U_t. Si $k_1, \ldots k_n \in D(V)$ alors

$$\lim_{t \to 0} \frac{1}{it}(\Gamma(U_t) - Id)(k_1 \circ \ldots \circ k_n) = \sum_{j=1}^{n} k_1 \circ \ldots \circ Vk_j \circ \ldots \circ k_n$$

par conséquent, $k_1 \circ \ldots \circ k_n$ est dans le domaine du générateur de $\Gamma(U_t)$.

DÉFINITION 31 – Soit A un opérateur sur K, la seconde quantification différentielle de A, notée $d\Gamma(A)$ est l'opérateur de domaine $\bigoplus_{n \in \mathbb{N}} D(A)^{\circ n}$ (somme algébrique) donné par

$$d\Gamma(A)(k_1 \circ \ldots \circ k_n) = \sum_{i=1}^{n} k_1 \circ \ldots \circ Ak_i \circ \ldots \circ k_n \qquad (63)$$

Lorsque V est auto-adjoint, $d\Gamma(V)$ est essentiellement auto-adjoint, en effet les vecteurs de la forme $k_1 \circ \ldots \circ k_n$ où les k_i sont analytiques pour V forment un ensemble total de vecteurs analytiques pour $d\Gamma(V)$. En particulier, si $(U_t)_{t \in \mathbb{R}}$ est un groupe unitaire fortement continu, le générateur du groupe $(\Gamma(U_t))_{t \in \mathbb{R}}$ est la fermeture de $d\Gamma(V)$. Dans la suite, si V est auto-adjoint, on désignera encore par $d\Gamma(V)$ la fermeture de $d\Gamma(V)$.

La seconde quantification différentielle de l'identité est appelée l'opérateur de nombre. C'est un opérateur auto-adjoint, le sous-espace $K^{\circ n}$ de $\Gamma(K)$ étant un espace propre de valeur propre n.

L'action des opérateurs de seconde quantification sur les vecteurs exponentiels a la forme suivante, soit $k \in D(A)$ (avec A auto-adjoint), alors k est dans le domaine de $d\Gamma(A)$ et

$$\Gamma(A)\xi(k) = \xi(Ak) \tag{64}$$
$$d\Gamma(A)\xi(k) = a^+_{Ak}\xi(k) \tag{65}$$

Il y a des relations de commutation entre les opérateurs de seconde quantification et les opérateurs de création et d'annihilation.

PROPOSITION 11 – Soient U un opérateur unitaire, A un opérateur auto-adjoint borné et $u \in K$ on a les relations suivantes

$$\Gamma(U)a^{\pm}_u\Gamma(U^{-1}) = a^{\pm}_{Uu} \tag{66}$$
$$[d\Gamma(A), a^{\pm}_u] = a^{\pm}_{Au} \tag{67}$$

sur le domaine $\Gamma_0(K)$.

Démonstration
Cela résulte d'un calcul simple utilisant les définitions de ces opérateurs.

On peut résumer les définitions des opérateurs de création, annihilation et seconde quantification en donnant les trois formules suivantes qui caractérisent leurs restrictions à l'espace $\Xi(K)$. Soient $u \in K$, A borné, pour tous $h, k \in K$ on a

$$< a^+_u\xi(k), \xi(h) > = < u, h >< \xi(k), \xi(h) > \tag{68}$$
$$< a^-_u\xi(k), \xi(h) > = < k, u >< \xi(k), \xi(h) > \tag{69}$$
$$< d\Gamma(A)\xi(k), \xi(h) > = < Ak, h >< \xi(k), \xi(h) > \tag{70}$$

4.4 Opérateurs de Weyl sur l'espace de Fock

Comme dans la proposition 4, on montre facilement que si A est borné, l'opérateur $d\Gamma(A) + a^+_u + a^-_u$ est essentiellement auto-adjoint. Nous allons introduire des opérateurs qui réalisent les groupes unitaires engendrés par les extensions auto-adjointes de ces opérateurs, en les définissant sur les vecteurs exponentiels.

PROPOSITION 12 – Soient U un opérateur unitaire sur K et $u \in K$ il existe un unique opérateur unitaire $W_{(U,u)}$ sur $\Gamma(K)$ tel que

$$\forall k \in K \ W_{(U,u)}\xi(k) = \xi(Uk + u)e^{-<Uk,u>-\frac{1}{2}|u|^2} \tag{71}$$

de plus, on a la relation

$$W_{(U,u)}W_{(V,v)} = W_{(UV,u+Uv)}e^{-i\Im m <u,Uv>} \tag{72}$$

Démonstration

Les vecteurs exponentiels forment un système libre, donc la formule (71) détermine un opérateur de domaine $\Xi(K)$. On vérifie facilement que c'est une isométrie, il suffit de voir que $< W_{(U,u)}\xi(k), W_{(U,u)}\xi(h) > = < \xi(k), \xi(h) > \ \forall k, h \in K$. Comme $\Xi(K)$ est dense dans $\Gamma(K)$, on peut étendre $W_{(U,u)}$ par continuité en une isométrie de $\Gamma(K)$. La formule (72) se vérifie sur les vecteurs exponentiels, et de là sur tout $\Gamma(K)$ par continuité. Elle entraine en particulier que les $W_{(U,u)}$ sont inversibles donc unitaires.

Remarquons que $W_{U,0} = \Gamma(U)$. La formule (72) montre que les opérateurs de la forme $W_{(U,u)}e^{i\theta}$ où U est unitaire, $u \in K$ et $\theta \in \mathbf{R}$, forment un groupe d'opérateurs unitaires. En fait si nous faisons de $U(K) \times K \times \mathbf{R}$ (où $U(K)$ est le groupe unitaire de K) un groupe avec le produit

$$(U, u, t) \star (U', u', t') = (UU', u + Uu', t + t' + \Im m < u, Uu' >)$$

alors

$$(U, u, t) \mapsto W_{(U,u)}e^{it}$$

est une représentation unitaire de ce groupe.

Nous allons en extraire certains sous-groupes à un paramètre et calculer les lois de leurs générateurs dans l'état vide.

Le premier exemple est celui des groupes $(W_{(Id,tu)})_{t\in\mathbf{R}}$. Son générateur s'obtient en calculant

$$\lim_{t\to 0} \frac{W_{(Id,tu)})\xi(h) - \xi(h)}{it} = \frac{1}{i}(a_u^+ - a_u^-)\xi(h)$$

Le générateur de ce groupe est donc $P_u = \frac{1}{i}(a_u^+ - a_u^-)$. Dans l'état vide, la variable aléatoire correspondante suit une loi de Gauss centrée de variance $|u|^2$, comme on peut le vérifier en calculant $< W_{(Id,tu)}\xi(0), \xi(0) > = e^{-|u|^2 \frac{t^2}{2}}$. D'après les relations de commutation (61) et (72), si on a un sous-espace vectoriel réel $\mathfrak{K}$ de K tel que $\forall u, v \in \mathfrak{K} < u, v > \in \mathbf{R}$ alors les opérateurs $(P_u)_{u\in\mathfrak{K}}$ (respectivement $(Q_u)_{u\in\mathfrak{K}}$) forment un processus classique dont la loi est celle d'un processus gaussien de covariance $\text{cov}(P_u, P_v) = < u, v >$.

Considérons maintenant des opérateurs de seconde quantification. Soient $U_t = e^{itA}$ un groupe fortement continu à un paramètre d'opérateurs unitaires, et $u \in K$. On sait que $(\Gamma(U_t))_{t\in\mathbf{R}}$ est un groupe unitaire fortement continu ainsi que son conjugué par l'opérateur $W_{(Id,u)}$

$$\left(W_{(Id,-u)}\Gamma(U_t)W_{(Id,u)} = W_{(U_t, U_t u - u)}e^{i\Im m < u, U_t u>}\right)_{t\in\mathbf{R}}$$

On peut caculer son générateur si $u \in D(A)$, c'est l'opérateur dont la restriction à $\Xi(K)$ est $d\Gamma(A) + a_{Au}^+ + a_{Au}^- + < u, Au >$, qui est essentiellement auto-adjoint. Pour calculer la loi de cette variable dans l'état vide, calculons sa transformée de Fourier.

On trouve

$$< W_{(Id,-u)} \Gamma(U_t) W_{(Id,u)} \Omega, \Omega > = < \Gamma(U_t) W_{(Id,u)} \Omega, W_{(Id,u)} \Omega >$$
$$= < \xi(U_t u) \, e^{-\frac{1}{2}|u|^2}, \xi(u) \, e^{-\frac{1}{2}|u|^2} >$$
$$= e^{<U_t u, u> - <u, u>}$$
$$= e^{\int_{-\infty}^{+\infty} (e^{itx} - 1) \, d\mu_u(x)}$$

où μ_u est la mesure spectrale de A associée au vecteur u.

D'après la formule de Lévy-Khintchine, cette loi est indéfiniment divisible, de mesure de Lévy μ_u.

Que devient le théorème de Stone von Neumann sur l'espace de Fock? Il s'agit de savoir si des opérateurs $\widetilde{W_u}$, $u \in K$ satifaisant aux relations de commutation

$$\widetilde{W_u} \widetilde{W_v} = \widetilde{W_{u+v}} e^{-i\Im m <u,v>} \tag{73}$$

se mettent sous la forme d'une somme de copies des opérateurs $W_{(I,u)}$. La réponse est oui si la dimension de K est finie (et la démonstration est essentiellement la même que celle donnée au chapitre 3 voir [48]). Par contre lorsque la dimension de K est infinie, il existe de nombreuses représentations irréductibles des relations (73) qui ne sont pas unitairement équivalentes à la représentation sur l'espace de Fock. Des exemples en sont fournis par le théorème de Shale [79]. Soit $\tau : K \to K$ une application **R**-linéaire qui préserve la forme symplectique $\Im m < u, v >$ sur K considéré comme espace réel, alors les opérateurs $W_{(I,\tau(u))}, u \in K$ forment une représentation des relations de commutation (73). Le théorème de Shale énonce que cette représentation est unitairement équivalente à la représentation $W_{(I,u)}$ (i.e. il existe U_τ unitaire tel que $U_\tau^* W_{(I,u)} U_\tau = W_{(I,\tau(u))} \; \forall u \in K$) si et seulement si $\tau^* \tau - Id$ est un opérateur de Hilbert-Schmidt. Ce théorème est à rapprocher du résultat classique suivant sur les mesures gaussiennes. Les lois de deux processus gaussiens linéaires $(X_u^1)_{u \in K}$ et $(X_u^2)_{u \in K}$ de covariances respectives q_1 et q_2 sont absolument continues si et seulement si $q_1 - q_2$ est de Hilbert-Schmidt (cf Neveu [57]). En fait la démonstration du théorème de Shale se ramène à cette propriété des mesures gaussiennes.

Le théorème de Shale est démontré dans [79] et [60].

4.5 Interprétations gaussienne et poissonnienne de l'espace de Fock

Lorsque K est de la forme $L^2_{\mathbf{C}}(E, \mathcal{E}, m)$ pour un espace mesurable $(E, \mathcal{E}, m)$ il y a deux interprétations probabilistes classiques de l'espace de Fock $\Gamma(K)$, comme espace L^2 d'un processus stochastique, que l'on va rappeler ci-dessous.

La première de ces interprétations est l'interprétation gaussienne. Donnons nous, sur un espace de probabilité $(\Omega, \mathcal{F}, P)$, une famille gaussienne X_B indexée par $\mathcal{E}$ de covariance $E(X_B X_C) = m(B \cap C)$. Le sous-espace fermé de $L^2(\Omega, \mathcal{F}, P)$ engendré par cette famille gaussienne est une famille gaussienne indexée par $L^2_{\mathbf{R}}(E, \mathcal{E}, m)$ de covariance $E(X_f X_g) = < f, g >$. Appelons $\mathcal{X}$ la tribu engendrée par les variables $X_B, B \in \mathcal{E}$.

PROPOSITION 13 – *L'application linéaire de* $\Xi(L^2_{\mathbb{C}}(E, \mathcal{E}, m))$ *dans* $L^2_{\mathbb{C}}(\Omega, \mathcal{X}, P)$ *qui à* $\xi(k)$ *associe la variable aléatoire*

$$e^{X_k - \frac{1}{2}|u|^2 + \frac{1}{2}|v|^2 + i<u,v>}$$

(avec $k = u + iv$*) se prolonge en un isomorphisme d'espaces de Hilbert de* $\Gamma(L^2_{\mathbb{C}}(E, \mathcal{E}, m))$ *dans* $L^2_{\mathbb{C}}(\Omega, \mathcal{X}, P)$.

Dans cet isomorphisme, l'opérateur de multiplication par la variable X_u *est transformé en l'opérateur* Q_u *sur* $\Gamma(L^2_{\mathbb{C}}(E, \mathcal{E}, m))$.

Démonstration

On vérifie que l'application en question est une isométrie grâce à la formule gaussienne:

$$E(e^{X_u + iX_v}) = e^{\frac{1}{2}<u,u> - <v,v> + i<u,v> + i<v,u>}$$

L'application se prolonge donc de façon unique par continuité à $\Gamma(L^2_{\mathbb{C}}(E, \mathcal{E}, m))$ en une isométrie. Pour voir qu'elle est surjective, il suffit de voir que les variables de la forme e^{X_k} forment une partie totale de $L^2_{\mathbb{C}}(\Omega, \mathcal{X}, P)$ (cf Neveu [57]).

Pour montrer l'assertion relative à Q_u on peut procéder de la manière suivante. On a $K = \mathbb{C}u + (\mathbb{C}u)^{\perp}$ d'où $\Gamma(K) = \Gamma(\mathbb{C}u) \otimes \Gamma((\mathbb{C}u)^{\perp})$. Pour les espaces L^2 cela correspond à la décomposition $L^2_{\mathbb{C}}(\Omega, \mathcal{X}, P) = L^2_{\mathbb{C}}(\Omega, \mathcal{X}_u, P) \otimes L^2_{\mathbb{C}}(\Omega, \mathcal{X}u^{\perp}, P)$ ($\mathcal{X}_u$ est la tribu engendrée par X_u et $\mathcal{X}u^{\perp}$ celle engendrée par les variables gaussiennes orthogonales à u). Dans cette décomposition on a $Q_u = Q_u \otimes Id$, et on utilise alors l'identification du chapitre 3 entre l'espace H et l'espace L^2 de la mesure de Gauss.

Les images des espaces $K^{\circ n}$ dans l'isométrie de la proposition 13 sont les chaos du processus gaussien, qui sont définis inductivement comme étant les sous-espaces C_n tels que $C_0 \oplus \ldots \oplus C_n$ est le sous-espace fermé engendré par les polynomes (complexes) de degrés $\leq n$ en les variables X_f.

La seconde interprétation de l'espace de Fock fait intervenir un processus de Poisson ponctuel de mesure caractéristique m. C'est par définition une famille $(M(B))_{B \in \mathcal{E}}$ de variables aléatoires telle que pour toute famille $B_1, \ldots, B_n$ d'éléments disjoints de $\mathcal{E}$ de mesures $m(B_i)$ finies, les variables $M(B_1), \ldots, M(B_n)$ sont des variables de Poisson indépendantes de paramètres $m(B_i)$. On peut trouver une version de ce processus définie sur l'espace canonique $(\mathfrak{M}, \mathcal{M})$ des mesures ponctuelles sur E (c'est à dire des mesures qui sont des sommes dénombrables de masses de Dirac). Un élément générique de cet espace est une mesure ponctuelle M. La loi P_m du processus ponctuel est donnée par sa fonction caractéristique

$$\int_{\mathfrak{M}} e^{\int_E f(x)dM(x)} dP_m(M) = e^{\int_E (e^{f(x)} - 1)dm(x)} \tag{74}$$

pour toute fonction f, telle que $e^f - 1$ soit m-intégrable.

PROPOSITION 14 – *Il existe un isomorphisme de* $\Gamma(L^2_{\mathbb{C}}(E, \mathcal{E}, m))$ *sur* $L^2_{\mathbb{C}}(\mathfrak{M}, \mathcal{M}, P_m)$ *tel que* $\xi(f)$ *pour* $f \in L^1 \cap L^2(E, \mathcal{E}, m))$ *corresponde à la variable* $\eta(f)$ *donnée par la formule*

$$\eta(f)(M) = \prod_{x \in \mathrm{supp}(M)} (1 + f(x))e^{- \int_E f(x)dm(x)} = e^{\int_E \mathrm{Log}(1+f(x))dM(x) - \int_E f(x)dm(x)} \tag{75}$$

Démonstration

Les variables de la forme $\eta(f)$ forment une partie totale dans $L^2(\mathfrak{M}, \mathcal{M}, P_m)$ (cf Neveu [58]), donc il suffit de montrer la formule $E(\eta(f)\eta(\bar{g})) = e^{\int_E f(x)\bar{g}(x)dm(x)}$ pour deux fonctions f et g de $L^1 \cap L^2(E, m)$, or d'après la formule caractéristique (74) on a

$$E(\eta(f)\eta(\bar{g})) = e^{\int_E (e^{\operatorname{Log}(1+f(x))(1+\bar{g}(x))}-1) \ dm(x) - \int_E f(x)+\bar{g}(x)dm(x)} = e^{\int_E f(x)\bar{g}(x)dm(x)}$$

Soit u une fonction réelle dans $L^2 \cap L^\infty(E, \mathcal{E}, m)$ la variable aléatoire correspondant à $a_u^+\xi(f)$ est

$$\lim_{t\to 0} \frac{\eta(f+tu)-\eta(f)}{t} = \Big(\int_E \frac{\cdot u(x)}{1+f(x)}dM(x) - \int_E u(x)dm(x)\Big)\eta(f)$$

La fonction u peut également être considérée comme un opérateur de multiplication sur $L^2(E, \mathcal{E}, m)$, et alors on a

$$d\Gamma(u)\eta(f) = a_{uf}^+\eta(f) = \Big[\int_E \frac{u(x)f(x)}{1+f(x)}dM(x) - \int_E u(x)f(x)dm(x)\Big]\eta(f)$$

On en déduit que l'opérateur $d\Gamma(u) + a_u^+ + a_u^- + \int_E u(x)dm(x)$, agissant sur l'espace de Fock s'interprète comme l'opérateur de multiplication par la variable aléatoire $\int_E u(x)dM(x)$. En particulier, prenant $u = 1_B$ avec $B \in \mathcal{E}$ et $m(B) < \infty$ on a $M(B) = d\Gamma(1_B) + a_{1_B}^+ + a_{1_B}^- + m(B)$.

4.6 Mouvement brownien et processus de Poisson

Nous allons maintenant nous intéresser plus particulièrement au cas où $K = L_{\mathbb{C}}^2(\mathbb{R}_+)$, et introduire ainsi les trois processus de Hudson et Parthasarathy. On définira au chapitre suivant des intégrales stochastiques par rapport à ces trois processus.

Il y a des identifications naturelles

$$L^2(\mathbb{R}_+)^{\circ n} \sim L^2(\mathbb{R}_+^n)_s \sim L^2(\Delta_n)$$

où $\Delta_n = \{(s_1, \ldots, s_n)|s_1 < \ldots < s_n\} \subset \mathbb{R}_+^n$ et $L_{\mathbb{C}}^2(\mathbb{R}_+^n)_s$ est le sous espace de $L_{\mathbb{C}}^2(\mathbb{R}_+^n)$ formé des fonction symétriques. Une fonction sur Δ_n est la restriction d'une unique fonction symétrique dans $\mathbb{R}_+^n$ ce qui donne la seconde identification, et pour la première, on peut identifier $h^{\circ n}$ avec la fonction $(s_1, \ldots, s_n) \mapsto \sqrt{n!}h(s_1)\ldots h(s_n)$.

D'après le paragraphe précédent, l'espace de Fock peut s'interpréter comme l'espace L^2 d'un processus gaussien, indexé par $L^2(\mathbb{R}_+)$. Un tel processus gaussien peut se réaliser à l'aide d'un mouvement brownien B par $X_f = \int f(s)dB_s$. Le mouvement brownien se retrouve par la formule $B_t = \int 1_{[0,t]}(s)dB_s$. Sur l'espace de Fock $\Gamma(L^2(\mathbb{R}_+))$, $Q_t = Q_{1_{[0,t]}}$ est l'opérateur de multiplication par la variable B_t. Le processus $(Q_t)_{t\in\mathbb{R}_+}$ est donc un mouvement brownien réel dans l'état vide. On dispose en fait de toute une famille de mouvements browniens, qui ne commutent pas entre eux, qui sont $P_t = P_{1_{[0,t]}}$, et toutes les combinaisons linéaires $\cos\theta\, Q_t + \sin\theta\, P_t$.

Nous allons maintenant interpréter les notions que nous avons introduites sur l'espace de Fock abstrait, au moyen du mouvement brownien B_t.

Soit $f \in L^2_{\mathbb{R}}(\mathbb{R}_+)$ l'opérateur Q_f est l'opérateur de multiplication par la variable aléatoire $\int_0^\infty f(s)dB_s$. L'opérateur a_f^- est l'opérateur de dérivation de Malliavin (gradient) dans la direction f et $L^2_{\mathbb{R}}(\mathbb{R}_+)$ est l'espace de Cameron-Martin du mouvement brownien.

Les chaos du mouvement brownien s'expriment à l'aide d'intégrales stochastiques itérées. Le n^e chaos est formé des intégrales stochastiques

$$\int \cdots \int_{\Delta_n} h(s_1, \ldots, s_n)dB_{s_1} \ldots dB_{s_n}$$

Le vecteur exponentiel $\xi(f)$ correspond à la variable aléatoire

$$e^{\int_0^\infty f(s)dB_s - \frac{1}{2}\int_0^\infty f^2(s)ds}$$

qui est la valeur terminale de la martingale exponentielle

$$t \mapsto e^{\int_0^t f(s)dB_s - \frac{1}{2}\int_0^t f^2(s)ds}$$

Passons maintenant à l'interprétation poissonienne de l'espace de Fock de $L^2_{\mathbb{C}}(\mathbb{R}_+)$. Un processus de Poisson ponctuel sur $\mathbb{R}_+$ ayant la mesure de Lebesgue comme mesure caractéristique donne un processus de Poisson $(N_t)_{t \in \mathbb{R}_+} = M([0,t])$ d'intensité 1. L'espace L^2 du processus de Poisson se décompose en chaos, le n^e chaos étant l'espace des intégrales stochastiques multiples

$$\int \cdots \int_{\Delta_n} f(s_1, \ldots, s_n)d\tilde{N}_{s_1} \ldots d\tilde{N}_{s_n}$$

$\tilde{N}_t = N_t - t$ étant le processus de Poisson compensé. Comme pour le mouvement brownien on obtient ainsi l'isomorphisme avec l'espace de Fock.
Sur l'espace de Fock, l'opérateur correspondant à la multiplication par N_t est $d\Gamma(1_{[0,t]}) + a^+_{1_{[0,t]}} + a^-_{1_{[0,t]}} + t$.
La variable N_t suit bien une loi de Poisson de paramètre t.
L'exponentielle $\xi(f)$ correspond à la variable $e^{\int_0^\infty \mathrm{Log}(1+f(s))dN_s - \int_0^\infty f(s)ds}$ qui est la valeur terminale de la martingale exponentielle $t \mapsto e^{\int_0^t \mathrm{Log}(1+f(s))dN_s - \int_0^t f(s)ds}$.

Nous avons maintenant rencontré les trois processus d'opérateurs fondamentaux de Hudson et Parthasarathy. Ce sont les processus a^-, a^+, a^0 sur l'espace $\Gamma(L^2_{\mathbb{C}}(\mathbb{R}_+))$ définis par

$$a_t^0 = d\Gamma(1_{[0,t]})$$
$$a_t^- = a^-_{1_{[0,t]}}$$
$$a_t^+ = a^+_{1_{[0,t]}}$$

Les processus $a_t^+ + a_t^-$ et $a_t^0 + a_t^+ + a_t^- + t$ sont des processus classiques, dans l'état vide leurs lois respectives sont celles d'un mouvement brownien, et d'un processus de Poisson de paramètre 1. La théorie classique du calcul stochastique par rapport à une martingale de carré intégrable montre que l'on dispose d'une notion d'intégrale stochastique par rapport à chacune des martingales $a_t^+ + a_t^-$ et $a_t^0 + a_t^+ + a_t^-$. Nous allons généraliser ce calcul stochastique en développant une théorie de l'intégrale stochastique par rapport aux trois processus non-commutatifs a_t^0, a_t^+, a_t^-. Dans cette théorie, les processus que nous chercherons à intégrer seront des processus d'opérateurs, c'est à dire des fonctions sur $\mathbf{R}_+$ à valeurs dans les opérateurs sur $\Gamma(L^2(\mathbf{R}_+))$.

4.7 Martingales normales et représentations chaotiques

En dehors des interprétations brownienne et poissonienne de l'espace de Fock $\Gamma(L_{\mathbf{C}}^2(\mathbf{R}_+))$ il existe d'autres représentations probabilistes, obtenues en considérant des martingales normales.

DÉFINITION 32 – Une martingale de carré intégrable M est dite normale si
$< M, M >_t = t$.
(Ici $< M, M >$ désigne le crochet de la martingale M, cf [21]).

Pour une telle martingale, On peut envoyer isométriquement l'espace de Fock $\Gamma(L_{\mathbf{C}}^2(\mathbf{R}_+))$ dans l'espace L^2 d'une telle martingale en posant

$$\phi(f_n) = \int_{\Delta_n} f_n(s_1, \ldots, s_n) dM_{s_1} \ldots dM_{s_n}$$

où $f_n \in L^2(\mathbf{R}_+)^{\circ n}$ est identifiée à une fonction de carré intégrable sur Δ_n.

DÉFINITION 33 – On dit que M possède la propriété de représentation chaotique si l'image de ϕ est L^2 tout entier.

Ainsi, dès que l'on dispose d'une martingale normale ayant la propriété de représentation chaotique, on a une interprétation probabiliste de l'espace de Fock. Alors que les propriétés de représentation chaotique du mouvement brownien et du processus de poisson compensé sont connues depuis longtemps, ce n'est que récemment que de nouvelles martingales possédant cette propriété ont été découvertes (voir [23]). Nous reviendrons plus en détails sur certains de ces exemples dans la suite.

Chapitre 5

Intégration stochastique non-commutative

5.1 Filtration et processus adaptés sur l'espace de Fock

L'espace de Hilbert $L^2_{\mathbb{C}}(\mathbb{R}_+)$ possède une structure de somme continue d'espaces de Hilbert, ce qui signifie que pour tout $t \in \mathbb{R}_+$ on a une décomposition orthogonale

$$L^2_{\mathbb{C}}(\mathbb{R}_+) = L^2_{\mathbb{C}}([0,t]) \oplus L^2_{\mathbb{C}}([t,\infty[)$$

et plus généralement si $t_1 < \ldots < t_n$

$$L^2_{\mathbb{C}}(\mathbb{R}_+) = L^2_{\mathbb{C}}([0,t_1]) \oplus L^2_{\mathbb{C}}([t_1,t_2]) \oplus \ldots \oplus L^2_{\mathbb{C}}([t_n,\infty[)$$

Au niveau des espaces de Fock, cela se traduit par une structure de *produit tensoriel continu*

$$\Gamma = \Gamma_{t_1]} \otimes \Gamma_{t_1,t_2} \otimes \ldots \otimes \Gamma_{[t_n}$$

où on a posé

$$\Gamma = \Gamma(L^2_{\mathbb{C}}(\mathbb{R}_+)), \ \Gamma_{t]} = \Gamma(L^2_{\mathbb{C}}([0,t])), \ \Gamma_{s,t} = \Gamma(L^2_{\mathbb{C}}([s,t])) \ \Gamma_{[t} = \Gamma(L^2_{\mathbb{C}}([t,+\infty]))$$

De même, pour les espaces engendrés par les vecteurs exponentiels on a, avec des notations semblables

$$\Xi = \Xi_{t_1]} \otimes \Xi_{[t_1,t_2]} \otimes \ldots \otimes \Xi_{[t_n}$$

(ici le produit tensoriel est pris au sens algébrique).

Dans la suite nous utiliserons souvent les notations $h_{t]} = h1_{[0,t]}$ et $h_{[t} = h1_{[t,+\infty[}$

Lorsqu'on utilise l'interprétation brownienne (ou poissonienne) de l'espace de Fock, en notant $\Omega_{s]}$ le vecteur vide de $\Gamma_{s]}$, l'espace L^2 engendré par les variables aléatoires mesurables par rapport à la tribu $\mathcal{F}_{[s,t]}$, engendrée par les accroissements du mouvement brownien (resp du processus de poisson) entre les instants s et t, est $\Omega_{s]} \otimes \Gamma_{[s,t]} \otimes \Omega_{[t}$. L'opérateur de multiplication correspondant à une variable X, $\mathcal{F}_{[s,t]}$-mesurable est de la forme $Id \otimes T_X \otimes Id$ sur la décomposition $\Gamma_{s]} \otimes \Gamma_{[s,t]} \otimes \Gamma_{[t}$ de l'espace de Fock. Ces remarques justifient la définition, donnée plus bas, des processus d'opérateurs adaptés sur l'espace de Fock.

Nous allons imiter la théorie des intégrales stochastiques par rapport à une martingale et définir des intégrales stochastiques par rapport aux trois processus $(a_t^+)_{t \in \mathbb{R}_+}$, $(a_t^-)_{t \in \mathbb{R}_+}$, et $(a_t^0)_{t \in \mathbb{R}_+}$. Comme on l'a vu au chapitre précédent, le mouvement brownien et le processus de Poisson s'expriment par des combinaisons linéaires de ces processus, il est donc naturel d'essayer de définir l'intégration stochastique par rapport à chacun de ces trois processus séparément. Les processus que nous allons intégrer seront des familles d'opérateurs $(X_t)_{t \in \mathbb{R}_+}$ sur l'espace Γ, adaptés au sens donné plus bas à ce mot. Pour définir ces intégrales stochastiques non-commutatives, la méthode usuelle (celle utilisée par Hudson et Parthasarathy dans leur article fondamental [40]) consiste à commencer par définir les intégrales stochastiques de processus simples (i.e. étagés) par la formule à laquelle on s'attend, puis à étendre par continuité l'application linéaire ainsi obtenue à une classe de processus plus générale, définie par une condition simple d'intégrabilité par raport à la mesure de Lebesgue, en utilisant une majoration de la norme de l'intégrale stochastique d'un processus simple. Cette méthode suppose que l'on sache approcher convenablement un processus "intégrable" par un processus simple, ce qui n'est pas si facile que cela. Dans ce cours, nous allons suivre une autre voie, inspirée de travaux de Belavkin et Lindsay. Nous allons commencer par définir les intégrales stochastiques de processus simples adaptés, puis obtenir une formule fondamentale qui caractérise ces intégrales stochastiques et qui a un sens pour des processus seulement intégrables. Ensuite nous utiliserons cette formule pour définir l'intégrale stochastique d'un processus adapté en général, la difficulté consistant à montrer l'existence de l'opérateur qui est caractérisé par la formule fondamentale. Pour cela nous allons utiliser de façon essentielle les propriétés de l'opérateur de divergence (définition 29).

Passons maintenant à la définition des processus adaptés.

DÉFINITION 34 – *Soit S un sous-espace dense de $L^2_{\mathbb{C}}(\mathbf{R}_+)$, stable par multiplication par les fonctions $1_{[s,t]}$ pour tous les $s < t \in \mathbf{R}_+$, une famille d'opérateurs $(X_t)_{t \in \mathbf{R}_+}$ sur Γ est dite $S-$adaptée (ou plus simplement adaptée si elle est S-adaptée pour un certain S) si pour tout $t \in \mathbf{R}_+$*

i) on a $D(X_t) \supset \Xi(S_{t]}) \otimes \Gamma_{[t}$, où $\Xi(S_{t]})$ est l'espace engendré par les vecteurs exponentiels de la forme $\xi(h1_{[0,t]}), h \in S$.

ii) X_t est de la forme $\widetilde{X_t} \otimes Id$ sur l'espace $\Xi(S_{t]}) \otimes \Gamma_{[t}$, (on note Φ_t l'ensemble des opérateurs de cette forme).

EXEMPLE – Les processus a^+, a^-, a^0 vérifient les relations $a^\varepsilon_t = \widetilde{a^\varepsilon_t} \otimes Id$ sur l'espace $\Xi_{t]} \otimes \Gamma_{[t}$, ils sont par conséquent $L^2_{\mathbb{C}}(\mathbf{R}_+)$-adaptés.

5.2 Intégration de processus simples

Dans ce paragraphe, on fixe S une partie de $L^2(\mathbf{R}_+)$ comme dans la définition 34.

DÉFINITION 35 – *Un processus adapté $(X_t)_{t \in \mathbf{R}_+}$ est dit simple s'il existe une suite de réels $0 = t_0 < t_1 < \ldots < t_n$ telle que $X_t = \sum_{j=0}^{n-1} 1_{[t_j, t_{j+1}[}(t)X^{(j)}$ où chaque $X^{(j)} \in \Phi_{t_j}$.*

Soient $(X_t)_{t \in \mathbf{R}_+}$ un processus adapté simple, $T_1 \leq T_2 \in \mathbf{R}_+ \cup \{\infty\}$ on définit trois opérateurs sur le domaine $\Xi(S_{T_2]}) \otimes \Gamma_{[T_2}$ ($\Xi(S)$ si $T_2 = \infty$) par la formule

$$\int_{T_1}^{T_2} X_s da^\varepsilon_s = \sum_{j=0}^{n-1} X^{(j)}(a^\varepsilon_{t_{j+1} \wedge T_2 \vee T_1} - a^\varepsilon_{t_j \wedge T_2 \vee T_1}) \tag{76}$$

où $\varepsilon = +, -$ ou 0.

DÉFINITION 36 – *Les opérateurs $\int_{T_1}^{T_2} X_s da^\varepsilon_s$ sont appelés les intégrales stochastiques de X par rapport aux trois processus a^ε, sur l'intervalle $[T_1, T_2]$.*

Vérifions immédiatement que la formule (76) définit bien un opérateur. Cette formule contenant des produits d'opérateurs non bornés, il faut vérifier qu'ils sont bien définis, or si $T_1 \leq s_1 \leq s_2 \leq T_2$ et X est dans Φ_{s_1}, sur la décomposition

$$\Xi(S_{T_2]}) \otimes \Gamma_{[T_2} = \Xi(S_{s_1]}) \otimes \Xi(S_{[s_1, T_2]}) \otimes \Gamma_{[T_2}$$

l'opérateur X est de la forme $\widetilde{X} \otimes Id \otimes Id$, et $a_{s_2}^\varepsilon - a_{s_1}^\varepsilon$ de la forme $Id \otimes (\widetilde{a_{s_2}^\varepsilon - a_{s_1}^\varepsilon}) \otimes Id$, par conséquent leur produit est bien défini sur ce domaine.

La proposition suivante va nous donner la caractérisation cherchée de l'intégrale stochastique d'un processus simple.

PROPOSITION 15 – Soit X un processus simple

$$\forall h, k \in S \quad < \int_{T_1}^{T_2} X_s \, da_s^\varepsilon(\xi(h)), \xi(k) > = \int_{T_1}^{T_2} < X_s(\xi(h)), \xi(k) > \kappa^\varepsilon(s) \, ds \quad (77)$$

avec $\kappa^- = h$, $\kappa^+ = \bar{k}$, $\kappa^0 = h\bar{k}$.

Démonstration

Par linéarité il suffit de le montrer pour un processus de la forme $X_t = X^1 1_{[t_1, t_2[}(t)$ avec $X^1 \in \Phi_{t_1}, T_1 \leq t_1 \leq t_2 \leq T_2$. Dans ce cas $\int_{T_1}^{T_2} X_s \, da_s^\varepsilon = X^1(a_{t_2}^\varepsilon - a_{t_1}^\varepsilon)$.

On a $X^1 = \widetilde{X^1} \otimes Id$ sur $\Xi(S_{t_1]}) \otimes \Xi_{[t_1}$ et $(a_{t_2}^\varepsilon - a_{t_1}^\varepsilon) = Id \otimes (\widetilde{a_{t_2}^\varepsilon - a_{t_1}^\varepsilon})$ sur le produit tensoriel $\Xi(S_{t_1]}) \otimes \Xi_{[t_1}$, donc $\int_{T_1}^{T_2} X_s \, da_s^\varepsilon = \widetilde{X^1} \otimes (\widetilde{a_{t_2}^\varepsilon - a_{t_1}^\varepsilon})$ et pour $h, k \in S$

$$< \int_{T_1}^{T_2} X_s \, da_s^\varepsilon(\xi(h)), \xi(k) > =$$

$$= < \widetilde{X^1}(\xi(h_{t_1]})) \otimes (\widetilde{a_{t_2}^\varepsilon - a_{t_1}^\varepsilon}) \xi(h_{[t_1}), \xi(k) >$$

$$= < \widetilde{X^1}(\xi(h_{t_1]})), \xi(k_{t_1]}) > < (\widetilde{a_{t_2}^\varepsilon - a_{t_1}^\varepsilon}) \xi(h_{[t_1}), \xi(k_{[t_1}) >$$

$$= < \widetilde{X^1}(\xi(h_{t_1]})), \xi(k_{t_1]}) > \int_{t_1}^{t_2} \kappa^\varepsilon(s) ds < \xi(h_{[t_1}), \xi(k_{[t_1}) >$$

d'après les formules (68)(69)(70)

$$= \int_{t_1}^{t_2} < X^1(\xi(h)), \xi(k) > \kappa^\varepsilon(s) \, ds$$

$$= \int_{T_1}^{T_2} < X_s(\xi(h)), \xi(k) > \kappa^\varepsilon(s) \, ds$$

Ce qui montre la proposition.

Comme $\Xi(S)$ est dense dans Γ la formule (77) détermine entièrement l'opérateur $\int_0^T X_s \, da_s^\varepsilon$ sur le domaine $\Xi(S)$.

Lorsque X est un processus qui n'est pas nécessairement simple ou adapté, si le membre de droite est bien défini pour tous les $h, k \in S$ (avec éventuellement $T_2 = +\infty$), comme les vecteurs exponentiels d'éléments de S forment une partie libre et totale dans $\Gamma(L_\mathbb{C}^2(\mathbb{R}_+))$ la formule (77) détermine au plus un opérateur $\int_{T_1}^{T_2} X_s \, da_s^\varepsilon$ sur le domaine $\Xi(S)$, et c'est cet opérateur que nous appellerons l'intégrale stochastique de X. Il reste à trouver un critère qui permet d'établir l'existence de cet opérateur.

Tout d'abord, nous allons énoncer quelques propriétés élémentaires de l'intégrale stochastique. Soit X (resp Y) un processus tel que

i) le membre de droite de (77) ait un sens pour tous les $h, k \in S$

ii) il existe un opérateur $\int_{T_1}^{T_2} X_s (\text{resp } Y_s) da_s^\varepsilon$ vérifiant (77)

alors les propriétés suivantes se vérifient immédiatement.

i) $\int_{T_1}^{T_2} X_s da_s^\varepsilon + \int_{T_1}^{T_2} Y_s da_s^\varepsilon = \int_{T_1}^{T_2} (X_s + Y_s) da_s^\varepsilon$

ii) $\int_{T_1}^{T_2} X_s da_s^\varepsilon + \int_{T_2}^{T_3} X_s da_s^\varepsilon = \int_{T_1}^{T_3} X_s da_s^\varepsilon$

iii) $\int_0^T X_s da_s^\varepsilon$ est de la forme $U \otimes Id$ sur $\Xi(S_{T]}) \otimes \Xi(S_{[T})$

iv) Posons $X_t^{S,T} = X_t$ pour $S < t < T$

$\qquad\qquad = 0$ sinon

alors

$$\int_0^\infty X_s^{S,T} da_s^\varepsilon = \int_S^T X_s da_s^\varepsilon$$

Il reste à trouver des conditions sur X pour qu'un opérateur vérifiant (77) existe. Pour cela, nous allons nous limiter aux intégrales de 0 à ∞ (grâce à iv) et considérer les trois cas $\varepsilon = -, 0, +$ séparément.

Le plus simple est celui de $\varepsilon = -$ car

$$\int_0^\infty < X_s(\xi(h)), \xi(k) > \kappa^-(s)\, ds =< \int_0^\infty h(s) X_s(\xi(h)) ds, \xi(k) >$$

par conséquent on peut poser

$$\left(\int_0^\infty X_s da_s^- \right) \xi(h)) = \int_0^\infty h(s) X_s(\xi(h)) ds \tag{78}$$

qui existe dès que la fonction de $\mathbf{R}_+$ dans Γ définie par $t \mapsto X_t(\xi(h))$ est faiblement mesurable, de carré intégrable.

Pour $\varepsilon = +$, on a

$$\int_0^\infty < X_s(\xi(h)), \xi(k) > \kappa^+(s)\, ds = \int_0^\infty < X_s(\xi(h)), k(s)\xi(k) > ds \tag{79}$$

Supposons de nouveau que la fonction $t \mapsto X_t(\xi(h))$ soit faiblement mesurable de carré intégrable. On peut la considérer comme un élément de $L_{\mathbf{C}}^2(\mathbf{R}_+) \otimes \Gamma(L_{\mathbf{C}}^2(\mathbf{R}_+))$, noté $X(\xi(h))$. De même, la fonction $t \mapsto k(t)\xi(k)$ définit un élément de $L_{\mathbf{C}}^2(\mathbf{R}_+) \otimes \Gamma(L_{\mathbf{C}}^2(\mathbf{R}_+))$. La définition (29) de l'opérateur gradient montre que cet élément est égal à $\nabla(\xi(k))$.

L'expression (79) vaut donc $< X(\xi(h)), \nabla(\xi(k)) >$ le produit scalaire étant pris dans $L_{\mathbf{C}}^2(\mathbf{R}_+) \otimes \Gamma(L_{\mathbf{C}}^2(\mathbf{R}_+))$. On voit donc qu'il existe un opérateur $\int_0^\infty X_s da_s^+(\xi(h))$ vérifiant (77) dès que $X(\xi(h))$ (i.e. la fonction $t \mapsto X_t(\xi(h))$ est dans le domaine de l'opérateur de divergence δ et alors, $\int_0^\infty X_s da_s^+(\xi(h)) = \delta(X(\xi(h)))$.

Un calcul du même type pour $\varepsilon = 0$ montre que $\int_0^\infty X_s da_s^0(\xi(h))$ existe dès que la fonction $t \mapsto h(t) X_t(\xi(h))$ considérée comme élément de $L_{\mathbf{C}}^2(\mathbf{R}_+) \otimes \Gamma(L_{\mathbf{C}}^2(\mathbf{R}_+))$ est dans le domaine de δ, et alors $\int_0^\infty X_s da_s^0(\xi(h)) = \delta(h X(\xi(h)))$.

Nous voyons donc que le problème d'existence de l'opérateur $\int X_s da_s^\varepsilon$ pour $\varepsilon = +, 0$ se ramène à celui de montrer qu'une fonction est dans le domaine de l'opérateur divergence. Nous allons étudier plus en détail les opérateurs gradient et divergence dans le paragraphe suivant. On sait en fait, depuis Gaveau et Trauber [36] que l'opérateur de divergence sur l'espace de Fock peut s'interpréter comme une intégrale stochastique, l'intégrale de Skorokhod [80]. Je renvoie à l'article de Nualart [59] pour une discussion plus détaillée de l'intégrale de Skorokhod et de ses liens avec l'opérateur de divergence.

5.3 Divergence et gradient sur l'espace $\Gamma(L_{\mathbf{C}}^2(\mathbf{R}_+))$

Rappelons que dans l'espace de Fock $\Gamma(L_{\mathbf{C}}^2(\mathbf{R}_+))$, le produit tensoriel symétrique $L_{\mathbf{C}}^2(\mathbf{R}_+)^{om}$ est isomorphe à l'espace $L_{\mathbf{C}}^2(\mathbf{R}_+^m)_s$ des fonctions symétriques de carré intégrable sur $\mathbf{R}_+^m$, le produit scalaire étant multiplié par $m!$. Un élément de $\Gamma(L_{\mathbf{C}}^2(\mathbf{R}_+))$ admet donc un développement convergent $a = \sum_{m=0}^{\infty} a_m$ où $a_m \in L_{\mathbf{C}}^2(\mathbf{R}_+^m)$ est une fonction symétrique.

Un élément u de $L_{\mathbf{C}}^2(\mathbf{R}_+) \otimes \Gamma(L_{\mathbf{C}}^2(\mathbf{R}_+))$ est donné par une fonction $t \mapsto u(t)$ de $\mathbf{R}_+$ dans Γ.

Si $t \in \mathbf{R}_+$ et $a = \sum_{m=0}^{\infty} a_m$ est un élément de $\Gamma(L_{\mathbf{C}}^2(\mathbf{R}_+))$ on pose $D_t a = \sum_{m=1}^{\infty} m a_m(., t)$.

PROPOSITION 16 – *Soit $a = \sum_{m=0}^{\infty} a_m \in \Gamma$, alors $a \in Dom(\nabla)$ si et seulement si $\int_0^{\infty} |D_t a|^2 dt < +\infty$ et alors $\nabla(a)$ est la fonction $t \mapsto D_t a \in L_{\mathbf{C}}^2(\mathbf{R}_+) \otimes \Gamma(L_{\mathbf{C}}^2(\mathbf{R}_+))$.*

Démonstration

Il suffit de montrer que pour toute $h \in L^2(\mathbf{R}_+)$ $\int_0^{\infty} \bar{h}(s) D_s a ds = a_h^-(a)$, or par définition de a_h^- on a $a_h^-(a) = \sum_{m=0}^{\infty} m \int_0^{\infty} a_m(., s) \bar{h}(s) ds$.

On note $\mathcal{L}^2$ la classe des fonctions $u \in L_{\mathbf{C}}^2(\mathbf{R}_+) \otimes \Gamma(L_{\mathbf{C}}^2(\mathbf{R}_+))$ telles que

$$\int_0^{\infty} \int_0^{\infty} |D_s u(t)|^2 ds dt < +\infty$$

Il est clair que l'espace $\mathcal{L}^2$ est dense dans $L_{\mathbf{C}}^2(\mathbf{R}_+) \otimes \Gamma(L_{\mathbf{C}}^2(\mathbf{R}_+))$ et que c'est un espace de Hilbert pour produit scalaire

$$< u, v >_{\mathcal{L}^2} = < u, v > + \int_0^{\infty} \int_0^{\infty} < D_s u(t), D_s v(t) > ds dt$$

Pour cette norme, l'espace vectoriel engendré par les éléments de la forme $h \otimes a$ où $h \in L_{\mathbf{C}}^2(\mathbf{R}_+)$ et a est dans la somme algébrique des espaces $L_{\mathbf{C}}^2(\mathbf{R}_+)^{ok}$, est dense dans $\mathcal{L}^2$.

PROPOSITION 17 – *On a $\mathcal{L}^2 \subset Dom(\delta)$ et pour tous $u, v \in \mathcal{L}^2$*

$$< \delta(u), \delta(v) > = < u, v > + \int_0^{\infty} \int_0^{\infty} < D_s u(t), D_t v(s) > ds dt \qquad (80)$$

Démonstration

Montrons d'abord la formule de la proposition pour u et v de la forme $u = h \otimes a$, $v = k \otimes b$ avec $h, k \in L_{\mathbf{C}}^2(\mathbf{R}_+)$ et $a, b \in \Gamma$ appartenant à la somme algébrique des sous-espaces $L_{\mathbf{C}}^2(\mathbf{R}_+)^{\otimes n}$. On a $D_s u(t) = h(t) D_s(a)$, et $D_t v(s) = k(s) D_t(b)$ et on calcule

$$
\begin{aligned}
< \delta(u), \delta(v) > &= < a_h^+(a), a_k^+(b) > \\
&= < a_k^- a_h^+(a), b > \\
&= < a_h^+ a_k^-(a).b > + < h, k >< a, b > \qquad \text{d'après (60)} \\
&= < a_k^-(a), a_h^-(b) > + < u, v >
\end{aligned}
$$

$$= \int_0^\infty \int_0^\infty < D_s u(t), D_t v(s) > ds dt + < u, v >$$

Cette égalité s'étend par linéarité au sous-espace engendré par les fonctions u comme ci-dessus, et en particulier pour u dans cet espace on a

$$|\delta(u)|^2 = |u|^2 + \int_0^\infty \int_0^\infty < D_s u(t), D_t u(s) > ds dt \leq |u|^2_{\mathcal{L}^2}$$

On en déduit la proposition par densité de ce sous-espace dans $\mathcal{L}^2$ et par le fait que δ est un opérateur fermé.

Nous avons vu au paragraphe précédent que l'intégrale stochastique par rapport à a^+ ou a^0 d'un processus adapté X_t appliquée à un vecteur exponentiel, si elle existe, s'exprime à l'aide de la divergence.
Si X est adapté, on a $X_t(\xi(h)) = X_t(\xi(h_{t]})) \otimes \xi(h_{[t})$ dans $\Gamma_{t]} \otimes \Xi_{[t}$, nous allons donc nous intéresser à l'appartenance au domaine de la divergence d'éléments de $L^2_{\mathbf{C}}(\mathbf{R}_+) \otimes \Gamma(L^2_{\mathbf{C}}(\mathbf{R}_+))$ de la forme $t \mapsto u(t) = x(t) \otimes \xi(h_{[t})$ avec $x(t) \in \Gamma_{t]}$.
On a $\nabla \xi(h) = h \otimes \xi(h)$ on en déduit que pour une telle fonction on a $D_s u(t) = h(s) x(t) \otimes \xi(h_{[t})$ si $s > t$ et $D_s u(t) = D_s x(t) \otimes \xi(h_{[t})$ si $s < t$. On en déduit le corollaire suivant. (Rappelons que pour une fonction u sur $\mathbf{R}_+$ on note $u_{s]}$ la fonction égale à u sur $[0, t]$ et à 0 sur $]t, +\infty[$.)

COROLLAIRE– On suppose que $(u(t) \equiv x(t) \otimes \xi(h_{[t}))_{t \in \mathbf{R}_+}$ et $(v(t) \equiv y(t) \otimes \xi(k_{[t}))_{t \in \mathbf{R}_+} \in L^2_{\mathbf{C}}(\mathbf{R}_+) \otimes \Gamma(L^2_{\mathbf{C}}(\mathbf{R}_+))$. Alors u, v sont dans le domaine de δ, et de plus

$$< \delta(u), \delta(v) > = \int_0^\infty < u(s), \delta(v_{s]}) > \bar{k}(s) ds + \int_0^\infty < \delta(u_{s]}), v(s) > h(s) ds$$
$$+ < u, v > \tag{81}$$

La divergence vérifie de plus l'inégalité suivante

$$|\delta(u)| \leq (|h| + \sqrt{|h|^2 + 1})(\int_0^\infty |u_s|^2)^{1/2} \tag{82}$$

Démonstration
Supposons tout d'abord que les fonctions x et y sont dans $\mathcal{L}^2$. Il s'ensuit que

$$\int_0^\infty \int_0^\infty |D_s u(t)|^2 ds dt = \int \int_{s<t} |D_s x(t)|^2 \otimes \xi(h_{[t})|^2 ds dt$$
$$+ \int \int_{s>t} |h(s) x(t) \otimes \xi(h_{[t})|^2 ds dt$$
$$\leq \int_0^\infty \int_0^\infty |D_s x(t)|^2 ds dt |\xi(h)|^2 + \int_0^\infty |u(t)|^2 dt \int_0^\infty |h(s)|^2 ds$$
$$< +\infty$$

donc la fonction u est dans $\mathcal{L}^2$ (ainsi que v pour la même raison). En appliquant la proposition précédente à ces deux fonctions on obtient

$$< \delta(u), \delta(v) > = < u, v > + \int_0^\infty \int_0^\infty < D_s u(t), D_t v(s) > ds dt$$

$$= < u, v > + \int_0^\infty \left[\int_0^t < D_s u(t), v(s) > ds \right] \bar{k}(t) dt$$

$$+ \int_0^\infty \left[\int_0^s h(t) < u(t), D_t v(s) > dt \right] ds$$

$$= < u, v > + \int_0^\infty < u(t), \delta(v_{t]}) > \bar{k}(t) dt$$

$$+ \int_0^\infty h(s) < \delta(u_{s]}), v(s) > ds$$

d'après la proposition 16, car δ est l'adjoint de ∇

ce qui montre la première identité dans le cas où $x, y \in \mathcal{L}^2$.

Nous allons en déduire l'inégalité (82) sur la norme de $\delta(u)$. D'après (81)

$$\left| \delta(u_{t]}) \right|^2 = \int_0^t |u(s)|^2 ds + 2 \int_0^t \Re \left(< u(s), h(s) \delta(u_{s]}) > \right) ds$$

d'où, en posant $\phi(t) = \sup_{s \leq t} |\delta(u_{s]})|$ et en appliquant l'inégalité de Cauchy-Schwarz

$$\phi(t)^2 \leq 2 \left(\int_0^t |h(s)|^2 ds \right)^{\frac{1}{2}} \left(\int_0^t | < u(s), \delta(u_{s]}) > |^2 ds \right)^{1/2} + \int_0^t |u(s)|^2 ds$$

$$\leq 2 \phi(t) |h| \left(\int_0^t |u(s)|^2 ds \right)^{\frac{1}{2}} + \int_0^t |u(s)|^2 ds$$

de cette dernière inégalité on tire

$$\left| \delta(u_{t]}) \right| \leq \left(|h| + \sqrt{|h|^2 + 1} \right) \left(\int_0^t |u(s)|^2 ds \right)^{1/2}$$

Soit maintenant une fonction $u(t) \equiv x(t) \otimes \xi(h_{[t})$ telle que pour tout t on ait $x(t) \in \Gamma_{t]}$ et $\int_0^\infty |u(t)|^2 dt < \infty$, on peut trouver une suite de fonctions $u^{(n)}$ de la forme $u^{(n)}(t) = x^{(n)}(t) \otimes \xi(h_{[t})$ telle que $x^{(n)} \in \mathcal{L}^2$ et $u^{(n)} \to_{n \to \infty} u$ dans $L^2_{\mathbf{C}}(\mathbf{R}_+) \otimes \Gamma(L^2_{\mathbf{C}}(\mathbf{R}_+))$ (par exemple on peut prendre pour $x^{(n)}(t)$ la projection de $x(t)$ sur l'espace engendré par les $n+1$ premiers chaos $\sum_{m \leq n} L^2_{\mathbf{C}}(\mathbf{R}_+)^{om}$). En utilisant l'inégalité (82) pour les $u^{(n)}$ et le fait que δ est un opérateur fermé, on voit que $u \in Dom(\delta)$ et les formules (81) et (82) s'obtiennent par passage à la limite.

5.4 Intégrale stochastique non-commutative et formule d'Itô

En appliquant le corollaire précédent à une fonction de la forme $u_t = X_t \xi(h)$, où X est un processus adapté, on obtient le résultat suivant, dû à Hudson et Parthasarathy.

THÉORÈME 7 – *Soient S une partie de $L^2_{\mathbf{C}}(\mathbf{R}_+)$ comme dans la définition (34), et X un processus S-adapté tel que $(t \mapsto \alpha^\varepsilon(t)X_t(\xi(h))) \in L^2_{\mathbf{C}}(\mathbf{R}_+) \otimes \Gamma(L^2_{\mathbf{C}}(\mathbf{R}_+))$ (où $\alpha^\pm(t) = 1, \alpha^0(t) = h(t)$) pour tout $h \in S$. Il existe un unique opérateur $\int_0^\infty X_s da^\varepsilon_s$ défini sur $\Xi(S)$ tel que pour tous $h, k \in S$ on ait*

$$< \int_0^\infty X_s da^\varepsilon_s(\xi(h)), \xi(k) >= \int_0^\infty < X_s\xi(h), \xi(k) > \kappa^\varepsilon(s)ds$$

avec $\kappa^+ = \bar{k}$, $\kappa^- = h$, $\kappa^0 = h\bar{k}$.
De plus on a

$$\int_0^\infty X_s da^-_s(\xi(h)) = \int_0^\infty h(s)X_s(\xi(h))ds \tag{83}$$

$$\int_0^\infty X_s da^0_s(\xi(h)) = \delta(hX(\xi(h))) \tag{84}$$

$$\int_0^\infty X_s da^+_s(\xi(h)) = \delta(X(\xi(h)) \tag{85}$$

Les intégrales stochastique vérifient les inégalités de norme

$$\left| \int_0^\infty X_s da^-_s \xi(h) \right| \le |h| \left(\int_0^\infty |X_s\xi(h)|^2 ds \right)^{\frac{1}{2}}$$

$$\left| \int_0^\infty X_s da^0_s \xi(h) \right| \le (|h| + \sqrt{|h|^2 + 1}) \left(\int_0^\infty |h(s)X_s\xi(h)|^2 ds \right)^{\frac{1}{2}} \tag{86}$$

$$\left| \int_0^\infty X_s da^+_s \xi(h) \right| \le (|h| + \sqrt{|h|^2 + 1}) \left(\int_0^\infty |X_s\xi(h)|^2 ds \right)^{\frac{1}{2}}$$

REMARQUES
1) Intégrales itérées.
Le processus $t \mapsto \int_0^t X_s da^\varepsilon$ est adapté, et d'autre part, les inégalités (86) montrent que pour $h \in S$, $t \mapsto \int_0^t X_s da^\varepsilon_s(\xi(h))$ est une application continue à valeurs dans l'espace de Fock. Il s'ensuit que l'on peut calculer l'intégrale stochastique itérée $\int_0^t (\int_0^s X_u da^\varepsilon_u) da^\eta_s$, où $\eta \in \{-, 0, +\}$ ainsi que des intégrales itérées d'ordre supérieur. Ceci sera très utile quand on cherchera à résoudre des équations différentielles stochastiques par la méthode de Picard au chapitre 6.
2) Adjoint d'une intégrale stochastique.
Soit X un processus S-adapté vérifiant les hypothèses du théorème (7) (X est dit ε-intégrable), supposons qu'il existe un processus S-adapté $*\varepsilon$-intégrable (où $*- = +, *0 = 0, *+ = -$) $X^\sharp$ tel que

$$\forall t \ge 0 \; \forall h, k \in S \; < X_t\xi(h), \xi(k) >=< \xi(h), X^\sharp_t\xi(k) >$$

(c'est à dire que $X^\sharp_t \subset X^*_t$) la formule (77) montre que

$$\forall \; h, k \in S \; < \int_0^\infty X_s da^\varepsilon_s(\xi(h)), \xi(k) >=< \xi(h), \int_0^\infty X^\sharp_s da^{*\varepsilon}_s(\xi(k)) > \tag{87}$$

Autrement dit on a $\int_0^\infty X^\sharp_s da^{*\varepsilon}_s \subset \left(\int_0^\infty X_s da^\varepsilon_s \right)^*$.

Nous avons ainsi réussi à définir l'intégrale stochastique de processus adaptés à l'aide de la formule (77). L'intégrale stochastique que nous obtenons est un opérateur défini sur un domaine exponentiel dense $\Xi(S)$. La première chose que nous allons faire est de vérifier que les intégrales stochastiques que nous venons de définir coïncident avec celles de la théorie classique. On va donc se placer dans l'interprétation brownienne de l'espace de Fock, et considérer le processus classique $a_t^+ + a_t^-$ comme un mouvement brownien B_t avec sa filtration $\mathcal{F}_t$. Soit X_t un processus adapté tel que chaque X_t est un opérateur de multiplication par une variable aléatoire (notée x_t) $\mathcal{F}_t$-mesurable. On va vérifier que l'intégrale stochastique non-commutative $\int_0^\infty X_s(da_s^+ + da_s^-)$ est la restriction au domaine $\Xi(S)$ de l'opérateur de multiplication par la variable aléatoire $\int_0^\infty x_s dB_s$. Pour cela, on utilise la formule fondamentale (77) et on calcule pour $h, k \in S$

$$E[\int_0^\infty x_s dB_s \xi(h)\overline{\xi(k)}]$$

Rappelons que dans l'interprétation brownienne $\xi(h)$ s'identifie à la variable aléatoire $\exp(\int_0^\infty h(s)dB_s - \frac{1}{2}\int_0^\infty h(s)^2 ds)$.

En particulier, $\xi(h)\overline{\xi(k)} = \xi(h + \bar{k})e^{<h,k>}$ le produit étant celui des variables aléatoires. D'autre part

$$\exp(\int_0^\infty h(s)dB_s - \frac{1}{2}\int_0^\infty h(s)^2 ds) = 1 + \int_0^\infty h(t)\xi(h_{t]})dB_t$$

on a donc, en utilisant la formule d'Ito

$$\int_0^\infty x_s dB_s \xi(h)\overline{\xi(k)} = e^{<h,k>}\int_0^\infty x_t\Big(\int_0^t \xi(h_{s]} + \bar{k}_{s]})(h(s) + \bar{k}(s))dB_s\Big)dB_t$$
$$+ e^{<h,k>}\int_0^\infty \Big(\int_0^t x_s dB_s\Big)\xi(h_{t]} + \bar{k}_{t]})(h(t) + \bar{k}(t))dB_t$$
$$+ e^{<h,k>}\int_0^\infty x_t\xi(h_{t]} + \bar{k}_{t]})(h(t) + \bar{k}(t))dt$$
$$+ \int_0^\infty x_s dB_s$$

En prenant l'espérance on trouve

$$E\Big[\int_0^\infty x_s dB_s \xi(h)\overline{\xi(k)}\Big] = E\Big[\int_0^\infty x_t\xi(h_{t]} + \bar{k}_{t]})(h(t) + \bar{k}(t))dt\Big]e^{<h,k>}$$
$$= \int_0^\infty < X_t\xi(h), \xi(k) > (h(t) + \bar{k}(t))dt$$
$$= < \int_0^\infty X_t(da_t^- + da_t^+)\xi(h), \xi(k) >$$

qui est bien ce qu'il fallait vérifier.

On démontre par un calcul analogue (bien qu'un peu plus fastidieux) que dans l'interprétation poissonnienne de l'espace de Fock, l'intégrale stochastique par rapport au processus de Poisson compensé s'obtient en prenant l'intégrale par rapport à $a^0 + a^- + a^+$.

Nous allons maintenant aborder l'un des aspects les plus remarquables du travail de Hudson et Parthasarathy qui est la généralisation de la formule d'Itô pour le mouvement brownien et le processus de Poisson aux intégrales stochastiques non-commutatives.

Une des façons possibles d'énoncer la formule d'Itô classique consiste à écrire le produit de deux intégrales stochastiques comme somme de trois termes, les deux premiers étant des intégrales stochastiques, et le troisième, le terme de correction d'Itô faisant intervenir le crochet des deux intégrales stochastiques. Plus précisément, si M et N sont deux martingales, H et K deux processus prévisibles intégrables, on a

$$\int_0^t H_s dM_s \int_0^t K_s dN_s = \int_0^t H_s \Big(\int_0^s K_u dN_u \Big) dM_s + \int_0^t \Big(\int_0^s H_u dM_u \Big) K_s dN_s$$
$$+ \int_0^t H_s K_s dM_s \cdot dN_s$$

Dans cette dernière expression on a noté $dM \cdot dN$ le crochet $d[M, N]$ afin d'éviter la confusion avec les commutateurs.

La formule d'Itô pour les intégrales stochastiques non-commutatives s'écrira formellement de la même façon

$$\int_0^t X_s da_s^\varepsilon \int_0^t Y_s da_s^\eta = \int_0^t X_s \Big(\int_0^s Y_u da_u^\eta \Big) da_s^\varepsilon + \int_0^t \Big(\int_0^s X_u da_u^\varepsilon \Big) Y_s da_s^\eta$$
$$+ \int_0^t X_s Y_s da_s^\varepsilon \cdot da_s^\eta$$

où $da_s^\varepsilon \cdot da_s^\eta$ est le "crochet" des processus a^ε et a^η. Nous calculerons ces crochets et nous verrons que suivant les valeurs de ε et η, le produit $da_s^\varepsilon \cdot da_s^\eta$ est égal à ds ou da_s^τ pour $\tau \in \{-, 0, +\}$.

L'intégrale stochastique étant un opérateur défini sur un domaine $\Xi(S)$, et ce domaine n'ayant aucune raison d'être stable, on ne peut pas multiplier (i.e. composer) deux intégrales stochastiques si bien que la formule ci-dessus n'a pas de sens en général. Pour tourner cette difficulté, on utilise l'identité

$$< \int_0^t X_s^* da_s^{*\varepsilon} \int_0^t Y_s da_s^\eta(\xi(h)), \xi(k) > = < \int_0^t Y_s da_s^\eta \xi(h), \int_0^t X_s da_s^\varepsilon \xi(k) >$$

justifiée formellement par la remarque 2 ci-dessus, et on en déduit que l'égalité formelle

$$< \int_0^t X_s^* da_s^{*\varepsilon} \int_0^t Y_s da_s^\eta \xi(h), \xi(k) > = \int_0^t < X_s^* \Big(\int_0^s Y_u da_u^\eta \Big) da_s^{*\varepsilon} \xi(h), \xi(k) >$$
$$+ \int_0^t < \Big(\int_0^s X_u^* da_u^{*\varepsilon} \Big) Y_s da_s^\eta \xi(h), \xi(k) >$$
$$+ < \int_0^t X_s^* Y_s da_s^{*\varepsilon} \cdot da_s^\eta \xi(h), \xi(k) >$$

est équivalente à l'égalité suivante qui, elle, a un sens pour $h, k \in S$

$$< \int_0^t Y_s da_s^\eta \xi(h), \int_0^t X_s da_s^\varepsilon \xi(k) > = \int_0^t < \int_0^s Y_u da_u^\eta \xi(h), X_s \xi(k) > \kappa^{*\varepsilon}(s) ds$$

$$+ \int_0^t < Y_s \xi(h), \int_0^s X_u da_u^\varepsilon \xi(k) > \kappa^\eta(s) ds$$

$$+ \int_0^t < Y_s \xi(h), X_s \xi(k) > \lambda^{\varepsilon\eta}(s) ds \quad (88)$$

où κ est défini comme dans (77) et $\lambda^{\varepsilon\eta} = \kappa^\tau$ si $da_s^{*\varepsilon} \cdot da_s^\eta = da_s^\tau$.

C'est cette dernière égalité que nous allons démontrer en calculant les valeurs des produits $da_s^\varepsilon \cdot da_s^\eta$ au moyen de la propriété fondamentale de la divergence (la formule (81)).

Commençons par un cas simple, $\varepsilon = \eta = -$ et calculons

$$< \int_0^t X_s da_s^- \xi(h), \int_0^t Y_s da_s^- \xi(k) > = \int_0^t \int_0^t h(s) \bar{k}(r) < X_s \xi(h), Y_r \xi(k) > ds dr$$

$$= \int_0^t < \int_0^s X_u da_u^- \xi(h), Y_s \xi(k) > \bar{k}(s) ds$$

$$+ \int_0^t h(s) < X_s \xi(h), \int_0^s Y_u da_u^- \xi(k) > ds$$

En rapprochant cette identité de (88) on obtient $da_s^+ \cdot da_s^- = 0$.

Les autres produits $da_s^\varepsilon . da_s^\eta$ se traitent de la même facon. Je vais expliciter le calcul de $da_s^- \cdot da_s^0$ et $da_s^- \cdot da_s^+$ et laisser les autres au lecteur.

D'après (81) on a

$$< \int_0^t X_s da_s^0 \xi(h), \int_0^t Y_s da_s^+ \xi(k) > = < \delta(hX(\xi(h))_{t]}), \delta(Y(\xi(k))_{t]}) >$$

$$= \int_0^t < \delta(hX(\xi(h))_{s]}), Y_s \xi(k) > h(s) ds$$

$$+ \int_0^t h(s) < X_s \xi(h), \delta(Y(\xi(k))_{s]}) > \bar{k}(s) ds$$

$$+ \int_0^t < h(s) X_s \xi(h), Y_s \xi(k) > ds$$

$$= \int_0^t < \int_0^s X_u da_u^0 \xi(h), Y_s \xi(k) > h(s) ds$$

$$+ \int_0^t h(s) < X_s \xi(h), \int_0^s Y_u da_u^+ \xi(k) > \bar{k}(s) ds$$

$$+ \int_0^t < h(s) X_s \xi(h), Y_s \xi(k) > ds$$

Par comparaison avec (88) nous obtenons donc les formules $da_s^- \cdot da_s^0 = da_s^-$ et en passant à l'adjoint $da_s^0 \cdot da_s^+ = da_s^+$.

Le calcul de $da^- \cdot da^+$ procède comme suit

$$< \int_0^t X_s da_s^+ \xi(h), \int_0^t Y_s da_s^+ \xi(k) > = < \delta(X(\xi(h))_{t]}), \delta(Y(\xi(k))_{t]}) >$$

$$= \int_0^t < \delta(X(\xi(h))_{s]}), Y_s \xi(k) > h(s) ds$$

$$+ \int_0^t < X_s \xi(h), \delta(Y(\xi(k))_{s]}) > \bar{k}(s) ds$$

$$+ \int_0^t < X_s \xi(h), Y_s \xi(k) > ds$$

$$= \int_0^t < \int_0^s X_u da_u^+ \xi(h), Y_s \xi(k) > h(s) ds$$

$$+ \int_0^t < X_s \xi(h), \int_0^s Y_u da_u^+ \xi(k) > \bar{k}(s) ds$$

$$+ \int_0^t < X_s \xi(h), Y_s \xi(k) > ds$$

et on obtient $da_s^- \cdot da_0^+ = ds$.

Finalement on trouve que la "table de multiplication d'Itô" s'écrit

$\cdot$	da_t^-	da_t^0	da_t^+
da_t^-	0	da_t^-	dt
da_t^0	0	da_t^0	da_t^+
da_t^+	0	0	0

On peut à partir de cette table retrouver les formules d'Itô pour le mouvement brownien et le processus de Poisson. Le mouvement brownien se réalise sur l'espace de Fock au moyen des opérateurs $B_t = a_t^+ + a_t^-$, et on a

$$d < B, B >_t = dB_t \cdot dB_t = (da_t^+ + da_t^-)^2 = dt$$

(ici $< B, B >$ est le crochet de la martingale B).

De même, le processus de Poisson compensé est $N_t = a_t^0 + a_t^+ + a_t^-$ et son crochet droit se calcule de la manière suivante

$$d[N, N]_t = dN_t^2 = (da_t^0 + da_t^+ + da_t^-)^2 = da_t^0 + da_t^+ + da_t^- + dt = dN_t + dt$$

On peut également incorporer à la table d'Itô un autre processus en plus des a^ε, qui est simplement le temps t. Il lui correspond les "intégrales stochastiques" $\int X_s ds$ qui sont des intégrales d'opérateurs au sens ordinaire. Il est immédiat de vérifier que la table de multiplication d'Itô se prolonge pour contenir dt, les produits $dt \cdot da_t^\varepsilon$ et $da_t^\varepsilon \cdot dt$ étant tous nuls.

A ce stade, se pose la question de savoir si on peut donner un sens directement à la formule

$$\int_0^t X_s da_s^\varepsilon \int_0^t Y_s da_s^\eta = \int_0^t X_s \left(\int_0^s Y_u da_u^\eta \right) da_s^\varepsilon + \int_0^t \left(\int_0^s X_u da_u^\varepsilon \right) Y_s da_s^\eta$$
$$+ \int_0^t X_s Y_s da_s^\varepsilon \cdot da_s^\eta$$

Par exemple, si le membre de droite de l'égalité est bien défini, on peut chercher à quelles conditions les opérateurs $\int_0^t X_s da_s^\varepsilon$ et $\int_0^t Y_s da_s^\eta$ sont prolongeables de façon à être composables.

La théorie des opérateurs de Maassen fournit une classe d'opérateurs qui possèdent un domaine dense et stable, ce qui fait qu'ils sont composables sur ce domaine. Ces opérateurs sont des sommes d'intégrales stochastiques multiples du type

$$\int f(s_1, \ldots s_n) da_{s_1}^{\varepsilon_1} \ldots da_{s_n}^{\varepsilon_n}$$

où f est une fonction déterministe. Sous des conditions convenable de croissance des fonctions f, ces opérateurs forment une classe stable par intégrale stochastique avec laquelle on peut exprimer directement la formule d'Itô. Je renvoie à [50] pour plus de détails sur ces opérateurs.

Pour terminer ce paragraphe, nous allons faire quelques calculs formels avec la table de multiplication d'Itô, pour arriver à un énoncé remarquable dû à Belavkin [6].

Soient X et Y deux processus adaptés qui s'écrivent sous la forme

$$X_t = \int_0^t H_s^- da_s^- + \int_0^t H_s^0 da_s^0 + \int_0^t H_s^+ da_s^+ + \int_0^t H_s ds$$

$$Y_t = \int_0^t K_s^- da_s^- + \int_0^t K_s^0 da_s^0 + \int_0^t K_s^+ da_s^+ + \int_0^t K_s ds$$

la formule d'Itô peut s'écrire de façon condensée

$$d(X_t Y_t) = X_t dY_t + (dX_t) Y_t + dX_t \cdot dY_t$$

avec $dX_t = \sum H_t^\varepsilon da_t^\varepsilon$, $dY_t = \sum K_t^\varepsilon da_t^\varepsilon$, les symboles da_t^ε commutant aux opérateurs K et H.

Introduisons les matrices à coefficients opérateurs suivantes

$$\mathbf{X}_t = \begin{pmatrix} X_t & 0 & 0 \\ 0 & X_t & 0 \\ 0 & 0 & X_t \end{pmatrix} \quad \mathbf{H}_t = \begin{pmatrix} 0 & H_t^- & H_t \\ 0 & H_t^0 & H_t^+ \\ 0 & 0 & 0 \end{pmatrix}$$

$$\mathbf{Y}_t = \begin{pmatrix} Y_t & 0 & 0 \\ 0 & Y_t & 0 \\ 0 & 0 & Y_t \end{pmatrix} \quad \mathbf{K}_t = \begin{pmatrix} 0 & K_t^- & K_t \\ 0 & K_t^0 & K_t^+ \\ 0 & 0 & 0 \end{pmatrix}$$

ainsi que la matrice

$$dA_t = \begin{pmatrix} 0 & 0 & 0 \\ da_t^- & da_t^0 & 0 \\ da_t & da_t^+ & 0 \end{pmatrix}$$

où on a posé $da_t^- = dt$. On a $dX_t = tr(H_t dA_t)$, et pour le processus adjoint $dX_t^* = tr(H_t^* dA_t)$ où l'involution $\star$ est définie par rapport à la forme quadratique de matrice $\begin{pmatrix} 0 & 0 & 1 \\ 0 & 1 & 0 \\ 1 & 0 & 0 \end{pmatrix}$ ce qui donne $H_t^* = \begin{pmatrix} 0 & (H_t^+)^* & (H_t)^* \\ 0 & (H_t^0)^* & (H_t^-)^* \\ 0 & 0 & 0 \end{pmatrix}$.

En utilisant ces matrices on peut réécrire la formule d'Itô

$$d(X_t Y_t) = tr\left[((X_t + H_t)(Y_t + K_t) - X_t Y_t) dA_t\right]$$

En particulier, on peut donner la forme fonctionnelle suivante de la formule d'Itô. Soit P un polynôme en une variable, alors

$$dP(X_t) = tr\left[(P(X_t + H_t) - P(X_t)) dA_t\right] \tag{89}$$

Cette formule est à rapprocher de la forme classique de la formule d'Itô pour le mouvement brownien. Si f est une fonction de classe C^2, alors

$$df(B_t) = f'(B_t) dB_t + \frac{1}{2} f''(B_t) dt$$

Cette formule se retrouve pour une fonction polynomiale f à partir de la formule de Belavkin (89) en considérant $X_t = a_t^+ + a_t^-$.

5.5 Exemples d'intégrales stochastiques

Commençons par l'exemple le plus simple. Pour toute $h \in L_{\mathbf{C}}^2(\mathbf{R}_+)$ on vérifie grâce à la formule (77) que
$\int_0^\infty h(s) da_s^+ = a_h^+$ et $\int_0^\infty \bar{h}(s) da_s^- = a_h^-$.
Si $h \in L^\infty(\mathbf{R}_+)$ est considéré comme un opérateur de multiplication sur $L_{\mathbf{C}}^2(\mathbf{R}_+)$ alors on a $d\Gamma(h_{t]}) = \int_0^t h(s) da_s^0$.
En prenant des produits de ces opérateurs et en appliquant la formule d'Itô on obtient

$$a_h^+ a_k^- = \int_0^\infty h(t)\left(\int_0^t \bar{k}(s) da_s^-\right) da_s^+ + \int_0^\infty \bar{k}(t)\left(\int_0^t h(s) da_s^+\right) da_s^-$$

$$a_k^- a_h^+ = \int_0^\infty h(t)\left(\int_0^t \bar{k}(s) da_s^-\right) da_s^+ + \int_0^\infty \bar{k}(t)\left(\int_0^t h(s) da_s^+\right) da_s^- + \int_0^\infty \bar{k}(t) h(t) dt$$

en particulier nous retrouvons la relation d'Heisenberg (60). Je laisse au lecteur le soin d'écrire les formules comprenant des intégrales par rapport à a^0, et de retrouver les relations (67).

Passons maintenant à une classe d'exemples plus générale, provenant de la théorie des martingales.

Soit M une martingale normale possédant la propriété de représentation chaotique (definitions 32 et 33). Nous pouvons donc identifier l'espace L^2 de cette martingale avec l'espace de Fock $\Gamma(L_{\mathbf{C}}^2(\mathbf{R}_+))$, ce que nous faisons dans la suite.

On suppose que la martingale M est dans L^4, ce qui entraine que son crochet est dans L^2 (cf [21]). Comme elle possède la propriété de représentation prévisible, il existe un processus prévisible ϕ tel que $[M, M]_t = t + \int_0^t \phi_s dM_s$. Nous allons noter par la même lettre ϕ le processus d'opérateurs de multiplication sur l'espace de Fock qu'il définit. Pour simplifier la discussion nous supposerons ϕ borné sur tout intervalle $[0, t]$.

Soit K un processus prévisible borné.

PROPOSITION 18 – Pour tout $t \in \mathbf{R}_+$, l'opérateur de multiplication par la variable $\int_0^t K_s dM_s$ est égal à $\int_0^t K_s\left(\phi_s da_s^0 + da_s^+ + da_s^-\right)$

Démonstration

Soient $u, v \in L^2(\mathbf{R}_+)$ et $H_t = \xi(v_{t]})$, $G_t = \xi(u_{t]})$. Quand on identifie l'espace de Fock avec l'espace L^2 de la martingale M au moyen des chaos, H et G sont deux martingales qui vérifient $H_t = 1 + \int_0^t v(s)H_s dM_s$ et $G_t = 1 + \int_0^t u(s)G_s dM_s$.

(remarque: la martingale M ayant un crochet oblique égal à $< M, M >_t = t$, les martingales H et G n'ont pas de sauts prévisibles, et donc si on en prend une version càdlàg, on a $H_{s-} = H_s$ presque surement pour tout s.) Si les processus H et G sont considérés comme des éléments de l'espace de Hilbert $L^2(\mathbf{R}_+) \otimes L^2(M)$ cette formule est la même que la formule usuelle $H_t = 1 + \int_0^t v(s)H_{s-}dM_s$).

Notons $\gamma_t = \int_0^t K_s dM_s$ Il faut montrer que

$$E[\gamma_t G_\infty \bar{H}_\infty] = \int_0^t < K_s\phi_s\xi(u), \xi(v) > u(s)\bar{v}(s)ds \qquad (90)$$

$$+ \int_0^t < K_s\xi(u), \xi(v) > (u(s) + \bar{v}(s))ds$$

On a

$$E[\gamma_t G_\infty \bar{H}_\infty] = E[\gamma_t G_t \bar{H}_t]e^{\int_t^\infty u(s)\bar{v}(s)ds} \qquad (91)$$

et

$$< K_t\phi_t\xi(u), \xi(v) > = < K_t\phi_t\xi(u_{t]}), \xi(v_{t]}) > e^{\int_t^\infty u(s)\bar{v}(s)ds}$$

En dérivant les deux égalités (90) et (91) par rapport à t, il suffit de vérifier que

$$E[\gamma_t G_t \bar{H}_t] = \int_0^t \Big(u(s)\bar{v}(s)E[\gamma_s G_s \bar{H}_s] + u(s) < K_s\phi_s\xi(u_{s]}), \xi(v_{s]}) > \bar{v}(s)$$

$$+ < K_s\xi(u_{s]}), \xi(v_{s]}) > (u(s) + \bar{v}(s))ds$$

Appliquons la formule d'Itô (pour la martingale M) au calcul de $\gamma_t G_t \bar{H}_t$. Tout d'abord on a

$$G_t \bar{H}_t = 1 + \int_0^t G_s d\bar{H}_s + \int_0^t \bar{H}_s dG_s + [G, \bar{H}]_t$$

$$= 1 + \int_0^t G_s \bar{H}_s \bar{v}(s) dM_s + \int_0^t u(s) G_s \bar{H}_s dM_s$$

$$+ \int_0^t u(s) G_s \bar{H}_s \bar{v}(s)(\phi_s dM_s + ds)$$

Si nous effectuons le produit avec γ_t, appliquons la formule d'Itô et prenons l'espérance, il suffit de garder les termes sans intégrale stochastique par rapport à M. On trouve

$$E[\gamma_t G_t \bar{H}_t] = E\Big[\int_0^t K_s G_s \bar{H}_s \bar{v}(s) ds + \int_0^t u(s) K_s G_s \bar{H}_s ds$$

$$+ \int_0^t u(s) G_s \bar{H}_s \bar{v}(s)(K_s \phi_s + \gamma_s) ds \Big]$$

Dans l'espace de Fock cela s'écrit

$$< \gamma_t \xi(u), \xi(v) > = \int_0^t < K_s \xi(u_{s]}), \xi(v_{s]}) > \bar{v}(s) ds$$

$$+ \int_0^t u(s) < K_s \xi(u_{s]}), \xi(v_{s]}) > ds$$

$$+ \int_0^t u(s) < (K_s \phi_s + \gamma_s) \xi(u_{s]}), \xi(v_{s]}) > \bar{v}(s) ds \Big]$$

c'est bien la formule qu'il fallait vérifier.

Nous allons appliquer ce dernier résultat à la construction des martingales d'Azéma, en suivant un article de Parthasarathy [61].

Les martingales d'Azéma sont des martingales de carré intégrable dont le crochet droit vérifie l'identité

$$d[X, X]_t = (c-1)X_{t-} dX_t + dt \tag{92}$$

M.Emery [26] a montré que pour toute valeur du paramètre réel c il existe une seule solution (en loi) à cette équation (avec $X_0 = 0$). Ce processus est un processus de Markov sur $\mathbb{R}$, et de plus lorsque $c \in [-1, +1]$ la martingale correspondante a la propriété de représentation chaotique. (A l'heure actuelle on ne sait pas si les martingales d'Azéma correspondant à une valeur de c en dehors de cet intervalle possèdent la propriété de représentation chaotique).

En utilisant la proposition précédente l'équation de structure (92) se transforme sur l'espace de Fock en l'équation

$$dX_t = (c-1)X_t da_t^0 + da_t^+ + da_t^- \tag{93}$$

(les variables aléatoires X_t et X_{t-} sont égales presque partout pour tout t). On va résoudre cette équation à l'aide d'intégrales stochastiques non-commutatives, et montrer que l'on obtient ainsi un processus classique, qui est une martingale solution de (92). Je vais me contenter d'esquisser les calculs et renvoyer à l'article [61] pour plus de détails.

L'équation (93) se résout par la méthode de variation des constantes. On commence par résoudre l'équation sans second membre $dJ_t^c = (c-1)J_t^c da_t^0$, dont la solution est $J_t^c = \Gamma(c1_{[0,t[} + 1_{[t,+\infty[})$. En cherchant la solution de l'équation complète sous la forme $X_t = J_t^c\big(\int_0^t x_s^+ da^+s + x_s^- da_s^-\big)$ et en appliquant la formule d'Itô on trouve $dX_t = (c-1)X_t da_t^0 + J_t^c(x_t^+ da_t^+ + x_t^- da_t^-) + (c-1)J_t^c x_t^+ da_t^-$ d'où les équations $J_t^c x_t^- = I$ et $cJ_t^c x_t^+ = I$. Ces équation se résolvent pour $c \neq 0$ et on obtient, pour $c \in [-1,1]\backslash\{0\}$,

$$X_t = J_t^c \int_0^t J_s^{c^{-1}}(da_s^- + c^{-1}da_s^+)$$

(Pour $c = 0$, la description de X se fait à l'aide de la notation de Guichardet pour l'espace de Fock, voir [61]).

Le processus X se décompose en somme de deux opérateurs adjoints l'un de l'autre $L_t^+ = J_t^c \int_0^t J_s^{c^{-1}} c^{-1} da_s^+$ et $L_t^- = J_t^c \int_0^t J_s^{c^{-1}} da_s^-$.
La formule d'Itô montre que pour tout $x \in \Xi(L_{\mathbf{C}}^2(\mathbf{R}_+))$

$$< L_t^+x, L_t^+x > -c < L_t^-x, L_t^-x > = < \int_0^t \Gamma(1_{[0,s[} + c^2 1_{[s,t[} 1_{[s,+\infty[})x, x > ds$$

On en déduit que si $c \in [-1,0[$ les opérateurs L_t^- et L_t^+ sont bornés sur Ξ et adjoints l'un de l'autre, donc X_t est auto-adjoint borné.

Montrons que le processus X est un processus classique. Fixons $s \in \mathbf{R}_+$, et posons $Y_t = Y_s$ pour $t \geq s$. D'après la formule d'Itô le commutateur $X_tY_t - X_tY_t$ vérifie l'équation $d(XY - YX)_t = (XY - YX)_t da_t^0$, et il est nul pour $t = s$ on en déduit par une application du lemme de Gronwall qu'il est nul pour $t \geq s$.

5.6 Extensions de l'intégrale stochastique

La première extension de la définition des intégrales stochastiques dont nous aurons besoin consiste à ajouter un espace de Hilbert à l'instant 0, c'est à dire à former le produit tensoriel $\mathcal{H}_0 \otimes \Gamma(L^2(\mathbf{R}_+))$ et à prolonger les opérateurs définis sur l'espace de Fock par l'identité sur $\mathcal{H}_0$. On peut ainsi définir l'intégrale stochastique de processus X d'opérateurs définis sur $D \otimes \Xi(S)$ où $D \subset \mathcal{H}_0$ est un domaine dense. La condition d'adaptation de X se traduit par une factorisation de X_t sur le produit $(D \otimes \Xi_{t]}) \otimes \Gamma_{[t}$. Les intégrales stochastiques sont données par des formules analogues à celles de de (77)

$$< \int_0^\infty X_s da_s^\varepsilon a \otimes \xi(h), b \otimes \xi(k) > = \int_0^\infty < X_s(a \otimes \xi(h)), b \otimes \xi(k) > \kappa^\varepsilon(s)ds$$

Il est facile en utilisant ce qui précède de vérifier que cette égalité détermine de façon unique un opérateur $\int_0^\infty X_s da_s^\varepsilon$ sur le domaine $D \otimes \Xi(S)$ dès que $\int_0^\infty |X_s a \otimes \xi(h)|^2 \alpha^\varepsilon(s)ds < +\infty$ pour tout $a \in D$ et $h \in S$.

Passons maintenant à la définition de l'intégrale stochastique en dimension supérieure c'est-à-dire sur l'espace de Fock multiple $\Gamma(L^2(\mathbf{R}_+) \otimes \mathbf{C}^d)$.

L'intégrale de Skorokhod étant un objet purement hilbertien, défini sur un espace de Fock quelconque, il est tentant d'essayer de formuler l'intégrale stochastique non-commutative en termes purement hilbertiens en se passant d'une identification de K avec $L^2(\mathbb{R}_+)$. C'est ce que nous allons faire dans ce paragraphe, en donnant une formule pour l'intégrale stochastique qui a l'avantage d'être simple et très générale, en particulier comme cette formule est la même sur tous les espaces de Fock, elle est valable sur un espace de Fock multiple $\Gamma(L^2(\mathbb{R}_+) \otimes \mathbb{C}^d)$.

Pour établir la formule en question, il faut remplacer la donnée d'un processus d'opérateurs X_t sur Γ par celle d'un seul opérateur sur un espace de Hilbert plus gros. C'est un peu l'analogue d'une idée commune en théorie générale des processus (classique!), au lieu de considérer un processus comme une famille X_t de variables aléatoires sur l'espace de probabilité Ω, on peut le considérer comme une variable aléatoire sur le gros espace $\mathbb{R}_+ \times \Omega$.

Commençons par expliciter cette formule pour l'intégrale par rapport au processus de création.

Soit $(X_t)_{t\in\mathbb{R}_+}$ un processus tel que pour tout $h \in S$, la fonction $t \mapsto X_t(\xi(h))$ soit de carré intégrable. C'est un élément de $L^2_{\mathbb{C}}(\mathbb{R}_+) \otimes \Gamma(L^2_{\mathbb{C}}(\mathbb{R}_+))$, donc $(X_t)_{t\in\mathbb{R}_+}$ définit un opérateur $X^+ : \Xi(L^2_{\mathbb{C}}(\mathbb{R}_+)) \to L^2_{\mathbb{C}}(\mathbb{R}_+) \otimes \Gamma(L^2_{\mathbb{C}}(\mathbb{R}_+))$, et on a $\int_0^\infty X_s da_s^+(\xi(h)) = \delta(X^+(\xi(h)))$ Nous voyons donc que l'intégrale stochastique de X n'est autre que la composition des opérateurs

$$\int_0^\infty X_s da_s^+ = \delta \circ X^+ \tag{94}$$

Passons maintenant au cas des intégrales d'annihilation. Soit X un processus, il définit un opérateur $X^- : L^2_{\mathbb{C}}(\mathbb{R}_+) \otimes \Xi(L^2_{\mathbb{C}}(\mathbb{R}_+)) \to L^2_{\mathbb{C}}(\mathbb{R}_+)$ en posant $X^-(k \otimes F) = \int_0^\infty k(s)X_s(F)ds$. La formule de définition de l'intégrale d'annihilation montre que

$$\int_0^\infty X_s da_s^-(\xi(h)) = \int_0^\infty X_s(\xi(h))h(s)ds = X^-(\nabla\xi(h))$$

L'intégrale stochastique est donc la composition

$$\int_0^\infty X_s da_s^- = X^- \circ \nabla \tag{95}$$

Pour l'intégrale de conservation, nous avons besoin d'une troisième interprétation d'un processus d'opérateurs. Au processus X on associe maintenant un opérateur X^0 de $L^2_{\mathbb{C}}(\mathbb{R}_+) \otimes \Xi(L^2_{\mathbb{C}}(\mathbb{R}_+))$ dans $L^2_{\mathbb{C}}(\mathbb{R}_+) \otimes \Gamma(L^2_{\mathbb{C}}(\mathbb{R}_+))$ en posant $X^0(k \otimes F) = t \mapsto k(t)X_t(F)$. L'opérateur X^0 ainsi obtenu commute avec les opérateurs de multiplication $h \otimes I : L^2_{\mathbb{C}}(\mathbb{R}_+) \otimes \Xi(L^2_{\mathbb{C}}(\mathbb{R}_+)) \to L^2_{\mathbb{C}}(\mathbb{R}_+) \otimes \Gamma(L^2_{\mathbb{C}}(\mathbb{R}_+))$. La formule (77) nous dit alors que $\int_0^\infty X_s da_s^0(\xi(h)) = \delta(X^0(\nabla\xi(h)))$ par conséquent, l'intégrale stochastique est cette fois la composition

$$\int_0^\infty X_s da_s^0 = \delta \circ X^0 \circ \nabla \tag{96}$$

Après ces considérations, nous pouvons donner l'extension multidimensionnelle de l'intégrale stochastique. On considère l'espace de Fock $\Gamma(L^2(\mathbb{R}_+) \otimes \mathbb{C}^d)$. Il est muni naturellement d'une structure de produit tensoriel continu, qui permet de définir la notion de processus adapté, et il peut s'interpréter comme l'espace L^2 d'un mouvement brownien (ou d'un processus de Poisson) de dimension d. Si on choisit une base e_j de $\mathbf{C}^d$, il y a un isomorphisme entre $\Gamma(L^2(\mathbf{R}_+) \otimes \mathbf{C}^d)$ et $\Gamma(L^2(\mathbf{R}_+))^{\otimes d}$ déduit de la proposition (10). En particulier, on a des opérateurs de création, annihilation et conservation sur la j^e composante du produit tensoriel

$$a_t^{+j} = a_{1_{[0,t]} \otimes e_j}^+ = Id \otimes \ldots \otimes a_t^+ \otimes \ldots \otimes Id,$$

$$a_t^{-j} = a_{1_{[0,t]} \otimes e_j}^- = Id \otimes \ldots \otimes a_t^- \otimes \ldots \otimes Id, \text{ et}$$

$$a_t^{0j} = d\Gamma(1_{[0,t]} \otimes E_{jj}) = Id \otimes \ldots \otimes a_t^0 \otimes \ldots \otimes Id.$$

Pour définir une intégrale de création, nous avons besoin d'un processus X^+ qui est un opérateur de $\Xi(L^2(\mathbf{R}_+) \otimes \mathbf{C}^d)$ dans $(L^2(\mathbf{R}_+) \otimes \mathbf{C}^d) \otimes \Gamma(L^2(\mathbf{R}_+) \otimes \mathbf{C}^d))$. En fixant une base e_j de $\mathbf{C}^d$, un tel opérateur est donné par une famille $(X_t^j)j = 1 \ldots d, t \in \mathbf{R}+$ d'opérateurs de $\Xi(L^2(\mathbf{R}_+) \otimes \mathbf{C}^d)$ dans $\Gamma(\Xi(L^2(\mathbf{R}_+) \otimes \mathbf{C}^d))$. L'intégrale de création de X^+ est alors la somme $\sum_j \int_0^\infty X_t^j da_t^{+j}$, chacun des termes de la somme étant défini à partir de la divergence. Une définition équivalente de cette intégrale stochastique consiste à écrire l'analogue de la formule (77)

$$< \int_0^\infty X_s da_s^+ \xi(h), \xi(k) >= \int_0^\infty < X_s \xi(h), k(s) \otimes \xi(k) > ds$$

où X_s est un opérateur de $\Xi(L^2(\mathbf{R}_+) \otimes \mathbf{C}^d)$ dans $\mathbf{C}^d \otimes \Gamma((L^2(\mathbf{R}_+) \otimes \mathbf{C}^d))$.

Comme dans le cas unidimensionnel, on vérifie que l'intégrale d'un processus adapté est toujours définie sur le domaine $\Xi(S \otimes \mathbf{C}^d)$, dès lors que les conditions $\int_0^\infty |X_s \xi(h \otimes e_j)|^2 ds < +\infty$ pour $h \in S$ sont remplies.

L'intégrale d'annihilation a une définition analogue, un processus X étant un opérateur de $L^2(\mathbf{R}_+) \otimes \mathbf{C}^d \otimes \Xi(L^2(\mathbf{R}_+) \otimes \mathbf{C}^d)$ dans $\Gamma(L^2(\mathbf{R}_+) \otimes \mathbf{C}^d)$, ou encore une famille X_s d'opérateurs de $\mathbf{C}^d \otimes \Xi(L^2(\mathbf{R}_+) \otimes \mathbf{C}^d)$ dans $\Gamma(L^2(\mathbf{R}_+) \otimes \mathbf{C}^d)$. La définition est cette fois contenue dans la formule

$$\int_0^\infty X_s da_s^- \xi(h) = \int_0^\infty X_s(h(s) \otimes \xi(h)) ds$$

En considèrant une base e_j de $\mathbf{C}^d$ nous pouvons encore décomposer cette intégrale en intégrales par rapport aux processus d'annihilation $a_t^{-j} = a_{1_{[0,t]} \otimes e_j}^-$.

La définition des intégrales de conservation va nous montrer que les processus a_t^{0j} introduits ci-dessus sont insuffisants pour décrire ces intégrales. Un processus intégrable par rapport au processus de conservation est un opérateur X^0 de $L^2(\mathbf{R}_+) \otimes \mathbf{C}^d \otimes \Xi(L^2(\mathbf{R}_+) \otimes \mathbf{C}^d)$ dans $L^2(\mathbf{R}_+) \otimes \mathbf{C}^d \otimes \Gamma(L^2(\mathbf{R}_+) \otimes \mathbf{C}^d)$ qui commute avec les opérateurs de multiplication $h \otimes I \otimes I$ sur $L^2(\mathbf{R}_+) \otimes \mathbf{C}^d \otimes \Xi(L^2(\mathbf{R}_+) \otimes \mathbf{C}^d)$. En introduisant une base e_j de $\mathbf{C}^d$ nous voyons qu'un tel processus se décompose en une famille $X_t^{jj'}$ d'opérateurs de $\Xi(L^2(\mathbf{R}_+) \otimes \mathbf{C}^d)$ dans $\Gamma(L^2(\mathbf{R}_+) \otimes \mathbf{C}^d)$. Son intégrale stochastique $\delta \circ X^0 \circ \nabla$ est alors une somme d'intégrales stochastiques $\sum_{jj'} \int_0^\infty X_t^{jj'} da_t^{0jj'}$ où les processus $a^{0jj'}$ sont définis par $a_t^{0jj'} = d\Gamma(1_{[0,t[} \otimes E_{jj'})$ ($E_{jj'}$ est l'opérateur qui envoie e_j sur $e_{j'}$ et les autres éléments de la base sur 0). Une formule analogue à (77) donne

$$< \int_0^\infty X_s^0 da_s^0 \xi(h), \xi(k) >= \int_0^\infty < X_s^0(h(s) \otimes \xi(h)), k(s) \otimes \xi(k) > ds$$

La formule d'Itô se prolonge au cas multidimensionnel nous avons cette fois d différentielles de processus de création da_t^{+j}, d d'annihilation da_t^{-j} et d^2 de conservation $da_t^{jj'}$. Il est plus agréable pour calculer la table de multiplication d'Itô d'adopter une notation plus intrinsèque en posant $a_t^{\pm u} = a_{1_{[0,t]} \otimes u}^{\pm}$ pour $u \in \mathbf{C}^d$ et $a_t^{0A} = d\Gamma(1_{[0,t]} \otimes A)$ pour $A \in M_d(\mathbf{C})$. On a alors

$$da_t^{-u} \cdot da_t^{+v} = <v, u> dt$$

$$da_t^{-u} \cdot da_t^{0A} = da^{-uA}$$

$$da_t^{0A} \cdot da_t^{+v} = da_t^{+Av}$$

(dans cette notation, u est un vecteur ligne et v un vecteur colonne).

Pour terminer signalons que la formule d'Itô sous la forme due à Belavkin (89) reste inchangée, si au lieu d'utiliser des matrices 3×3 on utilise des matrices $(d+2) \times (d+2)$. On considère des X et Y deux processus adaptés qui s'écrivent sous la forme

$$X_t = \sum_\varepsilon \int_0^t H_s^\varepsilon da_s^\varepsilon \qquad Y_t = \sum_\varepsilon \int_0^t K_s^\varepsilon da_s^\varepsilon$$

où ε parcourt les indices $-j$, $+j$, $0jj'$, $\cdot$ (avec toujours $da^\cdot = dt$). H_t^+ est un vecteur colonne d'opérateurs H_t^- un vecteur ligne et H_t^0 une matrice $d \times d$ avec lesquels on forme les matrices $(d+2) \times (d+2)$ à coefficients opérateurs

$$\mathbf{H}_t = \begin{pmatrix} 0 & H_t^- & H_t^{\cdot} \\ 0 & H_t^0 & H_t^+ \\ 0 & 0 & 0 \end{pmatrix} \mathbf{K}_t = \begin{pmatrix} 0 & K_t^- & K_t^{\cdot} \\ 0 & K_t^0 & K_t^+ \\ 0 & 0 & 0 \end{pmatrix}$$

$$d\mathbf{A}_t = \begin{pmatrix} 0 & 0 & 0 \\ da_t^- & da_t^0 & 0 \\ da_t^{\cdot} & da_t^+ & 0 \end{pmatrix}$$

Le processus adjoint de X satisfait toujours à $dX_t^* = tr(\mathbf{H}_t^\star d\mathbf{A}_t)$ où l'involution $\star$ est définie par rapport à la forme quadratique de matrice $\begin{pmatrix} 0 & 0 & 1 \\ 0 & Id & 0 \\ 1 & 0 & 0 \end{pmatrix}$, et la formule d'Itô est encore

$$d(X_t Y_t) = tr\left[((\mathbf{X}_t + \mathbf{H}_t)(\mathbf{Y}_t + \mathbf{K}_t) - \mathbf{X}_t \mathbf{Y}_t)d\mathbf{A}_t\right]$$

La forme fonctionnelle

$$dP(X_t) = tr\left[(P(\mathbf{X}_t + \mathbf{H}_t) - P(\mathbf{X}_t))d\mathbf{A}_t\right]$$

reste également valable.

Chapitre 6

Applications du calcul stochastique non-commutatif

6.1 Processus de Markov non-commutatif

Il existe une littérature importante consacrée aux applications du calcul stochastique non-commutatif à la physique quantique, qu'il ne m'est pas possible d'aborder ici faute de compétence. Les lecteurs intéressés par ces applications pourront consulter les volumes des rencontres de probabilités quantiques édités par L.Accardi et W.von Waldenfels [2] ainsi que les références qu'ils contiennent. Mentionnons juste que le calcul stochastique non-commutatif sert à construire des "évolutions stochastiques" qui décrivent le comportement d'un système quantique ouvert, en interaction avec une "source de bruit quantique" représentée par un espace de Fock. L'évolution dans le temps d'un système quantique fermé est gouvernée par un opérateur auto-adjoint, l'hamiltonien, et le groupe unitaire qu'il engendre. Lorsque le système S en considération est en interaction avec un autre système (appelé "l'extérieur") le système total formé par la réunion de S et de l'extérieur est fermé et son évolution est donc gouvernée par un hamiltonien. Lorsqu'on ne s'intéresse qu'au système lui-même, l'interaction avec l'extérieur fait que son évolution n'est plus gouvernée par un hamiltonien mais par un opérateur "dissipatif" (je renvoie au livre de Davies [18] pour une discussion précise de ces notions). Il est apparu que la théorie de l'évolution des systèmes quantiques ouverts possède une analogie formelle avec la théorie classique des processus de Markov, l'opérateur dissipatif qui décrit l'évolution du système ouvert étant un analogue non-commutatif du générateur d'un semi-groupe markovien.

Je vais essayer de donner quelques rudiments de cette théorie des processus de Markov non-commutatifs. Nous allons quitter le terrain de la physique quantique que nous n'avons fait qu'effleurer pour nous concentrer sur l'aspect probabiliste de la question.

Commençons par quelques rappels sur la théorie élémentaire des processus de Markov.

Soient E un espace lusinien (pour fixer les idées) et $(P_t)_{t \in \mathbb{R}_+}$ un semi-groupe de noyaux markoviens sur E. On sait qu'il est possible de construire un processus de Markov associé, qui consiste en la donnée d'un espace de probabilité $(\Omega, \mathcal{F}, P)$, d'une filtration $(\mathcal{F}_t)_{t \in \mathbb{R}_+}$, et d'une famille de variables aléatoires $X_t : \Omega \to E$, $t \in \mathbb{R}_+$ adaptée à la filtration $\mathcal{F}_t$ tels que

$$\forall t_1 < \ldots < t_n \ \forall f \text{ borélienne bornée sur } E^n$$
$$E[f(X_{t_1}, \ldots, X_{t_n})] =$$
$$\int_{E^{n+1}} f(x_1, \ldots, x_n) \mu(dx_0) P_{t_1}(x_0, dx_1) \ldots P_{t_n - t_{n-1}}(x_{n-1}, dx_n) \tag{97}$$

(le processus X est de loi initiale μ).

Un procédé usuel pour obtenir un tel processus consiste à utiliser le théorème de Kolmogorov pour construire la probabilité P sur l'espace canonique $\Omega = E^{\mathbb{R}+}$, muni des applications coordonnées X_t, comme limite projective des lois marginales de rang fini données par la formule (97).

La propriété de Markov simple du processus X se traduit par l'identité

$$E[f(X_t)|\mathcal{F}_s] = P_{t-s}f(X_s) \qquad (98)$$

vérifiée pour tout couple $s < t \in \mathbb{R}_+$ et toute fonction f borélienne bornée sur E.

Plusieurs auteurs ont proposé des définitions de la notion de processus de Markov non-commutatif (voir par exemple [1], [29], [9], [74], [84]). Sans entrer dans les détails, nous allons examiner les principes qui sont à la base de ces définitions.

Suivant les principes des "probabilités quantiques", nous allons commencer par traduire la situation précédente en termes algébriques. Le semi-groupe markovien P_t détermine un semi-groupe d'applications linéaires T_t sur l'algèbre $B(E)$ des fonctions boréliennes bornées sur E définies par $T_t f(x) = \int_E f(y) P_t(x, dy)$. Les applications T_t sont positives, (l'image d'une fonction positive est positive) et préservent l'identité car P_t est markovien. Réciproquement, un semi-groupe d'applications linéaires positives sur $B(E)$, préservant l'identité, provient d'un unique semi-groupe de noyaux markoviens (cf [22], on utilise le fait que E est lusinien), les données de P_t et de T_t sont donc équivalentes.

Le processus X_t définit une famille de morphismes d'algèbres $j_t : B(E) \to L^\infty(\Omega, \mathcal{F}, P)$ en posant $j_t(f) = f \circ X_t$. Ces morphismes déterminent le processus X car si Y est un autre processus tel que $f \circ Y_t = f \circ X_t$ pour tout fonction borélienne bornée, alors $X_t = Y_t$.

Si E possède une mesure m naturellement associée au semi-groupe P_t, (par exemple la mesure de comptage si E est dénombrable, ou une mesure m par rapport à laquelle les noyaux $(P_t)_{t>0}$ sont absolument continus), on peut définir les j_t sur l'algèbre $L^\infty(E, m)$, et le processus X est encore déterminé à une modification près par les morphismes j_t.

L'espérance conditionnelle sur la tribu $\mathcal{F}_t$ définit une projection

$$E_t : L^\infty(\Omega, \mathcal{F}, P) \to L^\infty(\Omega, \mathcal{F}_t, P)$$

et la propriété de Markov (98) devient alors

$$E_s \circ j_t = j_s \circ T_{t-s}$$

Décrivons maintenant des analogues non-commutatifs des objets précédents.

La donnée de base, qui est celle de l'espace E et du semi-groupe P_t est remplacée par celle d'une algèbre involutive à unité $\mathcal{A}$, qui joue le role d'analogue de l'algèbre $B(E)$ (ou de $L^\infty(E, m)$), et d'un semi-groupe T_t d'applications complètement positives de $\mathcal{A}$, préservant l'identité.

Expliquons ces termes.

Une algèbre involutive est une algèbre sur $\mathbb{C}$, munie d'une involution $a \mapsto a^*$ qui est une application antilinéaire, telle que $(ab)^* = b^* a^*$ (cf Dixmier [24]). Dans la pratique, $\mathcal{A}$ sera le plus souvent une C^*−algèbre.

DÉFINITION 37 − Une C^*-algèbre est une algèbre involutive munie d'une norme qui en fait un espace de Banach telle que
i) $\forall x, y \in \mathcal{A}\ |xy| \leq |x||y|$ (c'est une algèbre de Banach)
ii) $|x^*| = |x|$ et $|xx^*| = |x|^2$

Un exemple immédiat est celui de $\mathcal{B}(H)$ (opérateurs bornés sur un Hilbert H) avec l'involution donnée par le passage à l'adjoint.

Un autre exemple est celui de l'algèbre $C_0(X)$ (fonctions complexes continues nulles à l'infini sur un espace topologique localement compact X, l'involution étant la conjugaison). En fait, d'après un célèbre théorème de Gelfand, toute $C^*−$algèbre commutative est isomorphe à une algèbre de ce type, l'espace X étant homéomorphe à l'ensemble des caractères de la $C^*−$algèbre.

Un élément a d'une $C^*−$algèbre est dit hermitien si $a^* = a$, et positif s'il est de la forme $a = x^*x$ pour un $x \in \mathcal{A}$.

Soit $\mathcal{A}$ une $C^*−$algèbre avec unité, un état sur $\mathcal{A}$ est une application linéaire φ continue, positive (elle envoie les éléments positifs de $\mathcal{A}$ dans $\mathbb{R}_+$), telle que $\varphi(1) = 1$. C'est l'analogue d'une mesure de probabilité, et les deux notions sont équivalentes lorsque $\mathcal{A}$ est commutative, d'après le théorème de Riesz.

Soient $\mathcal{A}$ et $\mathcal{B}$ des $C^*−$algèbres, une application linéaire T de $\mathcal{A}$ dans $\mathcal{B}$ est dite positive si l'image par T d'un élément positif de $\mathcal{A}$ est un élément positif de $\mathcal{B}$. La notion d'application positive entre deux $C^*−$algèbres a le défaut de ne pas être stable par produit tensoriel, si T est positive, il se peut que l'application $T \otimes Id : \mathcal{A} \otimes M_n(\mathbb{C}) \to \mathcal{B} \otimes M_n(\mathbb{C})$ ne soit pas positive. Pour cette raison, l'analogue non-commutatif de la notion d'application positive que nous considèrerons est celle d'application complètement positive.

DÉFINITION 38 − On dit que l'application linéaire $T : \mathcal{A} \to \mathcal{B}$ est complètement positive si pour tout $n \geq 0$ l'application linéaire $T \otimes Id : \mathcal{A} \otimes M_n(\mathbb{C}) \to \mathcal{B} \otimes M_n(\mathbb{C})$ est positive

Si $\mathcal{A}$ est commutative, les notions d'application positive et complètement positive coïncident.

DÉFINITION 39 − Soient $\mathcal{A}$ une $C^*−$algèbre et ω un état sur $\mathcal{A}$, une espérance conditionnelle sur $\mathcal{A}$ est une application linéaire complètement positive $E : \mathcal{A} \to \mathcal{A}$, préservant l'identité, telle que
$\omega \circ E = \omega$, et $\forall x, y \in \mathcal{A}\ E(xE(y)) = E(x)E(y)$.

L'image d'une espérance conditionnelle est une sous-algèbre de $\mathcal{A}$ dont E laisse les éléments invariants. Lorsque $\mathcal{A} = L^\infty(\Omega, \mathcal{F}, P)$ et $\mathcal{G} \subset \mathcal{F}$ l'espérance conditionnelle usuelle sur $L^\infty(\Omega, \mathcal{G}, P)$ est une espérance conditionnelle au sens de la définition précédente.

On suppose donnés une C^*-algèbre $\mathcal{A}$ avec une identité, et un semi-groupe T_t d'applications complètement positives sur $\mathcal{A}$, préservant l'identité. Le semi-groupe T_t joue le role de semi-groupe markovien. On peut alors se poser le problème construire un "processus de Markov non-commutatif" correspondant.

Un tel processus sera la donnée d'une dilatation du semi-groupe T_t.

DÉFINITION 40 − Une dilatation du semi-groupe T_t est la donnée de $(\mathcal{W}, \omega, \mathcal{W}_t, E_t, j_t)$ où
i) $\mathcal{W}$ est une C^*-algèbre à unité munie d'un état ω
ii) $(\mathcal{W}_t)_{t \in \mathbb{R}_+}$ est une famille croissante de sous-algèbres de $\mathcal{W}$.

iii) Pour tout $t \in \mathbb{R}_+$, E_t *est une espérance conditionnelle d'image* $\mathcal{W}_t$, *et* $\forall s, t \in \mathbb{R}_+$, $E_t E_s = E_{s \wedge t}$

iv) Les $j_t : \mathcal{A} \to \mathcal{W}_t$ *sont des morphismes qui préservent l'identité et vérifient la propriété de Markov*

$$E_s \circ j_t = j_s \circ T_{t-s}$$

La définition d'une dilatation donnée ci-dessus n'est pas la seule possible, on peut demander par exemple que $\mathcal{A}$ et $\mathcal{W}$ soient des algèbres de von Neumann, avec ω, E_t, j_t normaux (cf Dixmier [24]), on peut aussi imposer des conditions moins fortes sur les E_t (comme dans Bhat et Parthasarathy [9]). Je renvoie aux articles cités pour des variantes possibles de cette définition, ainsi que pour les théorèmes d'existence correspondants.

Nous allons construire des dilatations de semi-groupes complètement positifs sur $\mathcal{B}(H_0)$ vérifiant certaines conditions analytiques en résolvant des équations différentielles stochastiques non-commutatives. Ces semi-groupes sont des analogues des semi-groupes de diffusions sur des variétés. Ensuite nous appliquerons ces idées à la construction de certains processus de Markov classiques. Le principe de ces constructions est le suivant. On se donne un espace E, avec une mesure m, et un semi-groupe markovien P_t sur E, le semi-groupe associé sur $L^\infty(E)$ étant T_t. Le semi-groupe T_t est étendu en un semi-groupe complètement positif de $\mathcal{B}(L^2(E, m))$, dont on construit une dilatation sur $\mathcal{W} = \mathcal{B}(H)$ telle que les algèbres $j_t(L^\infty(E))$ commutent. Soit $\mathcal{A}$ l'algèbre de von Neumann commutative engendrée par les $j_t(L^\infty(E))$, d'après le théorème de structure des algèbres de von Neumann commutatives (cf [24]), il existe un espace de probabilité $(\Omega, \mathcal{F}, P)$ et des variables aléatoires $X_t : \Omega \to E$ telles que $f \circ X_t = j_t(f)$ pour $f \in L^\infty(E)$. Les variables X_t forment alors un processus de Markov sur E de semi-groupe P_t.

Avant de réaliser ce programme, nous allons montrer au paragraphe suivant comment la marche de Bernoulli quantique du chapitre 1 peut être considérée comme un processus de Markov non-commutatif.

6.2 Un autre construction de la marche de Bernoulli quantique

Dans ce paragraphe nous allons donner un exemple explicite de dilatation d'un semi-groupe d'applications complètement positives en temps discret. Ce type de chaîne de Markov non-commutative est l'analogue d'une marche aléatoire sur un groupe. Nous verrons qu'il permet d'interpréter naturellement la marche de Bernoulli quantique du chapitre 1 comme une chaîne de Markov non-commutative, dont le processus de spin serait la "partie radiale".

Ce paragraphe est une parenthèse, il ne servira pas pour la suite du cours. Le lecteur peu familier avec la théorie des groupes peut passer directement au paragraphe 6.3. Commençons par quelques rappels de théorie des groupes. Je renvoie à [24] et [25] pour ce qui concerne les algèbres de groupes, et à [14] et [83] pour les représentations des groupes compacts en général et de SU(2) en particulier.

Soit G un groupe compact, muni de sa mesure de Haar normalisée. Il agit sur l'espace $L^2(G)$ par multiplication à gauche, i.e. chaque élément g du groupe définit un opérateur unitaire λ_g sur $L^2(G)$ par $\lambda_g f(h) = f(g^{-1}h)$. Appelons $vN(G)$ l'algèbre de von Neumann engendrée par ces opérateurs. Soit ν un état normal sur cette algèbre, la fonction ϕ sur G définie par $\phi(g) = \nu(\lambda_g)$ est une fonction de type positif, continue.

Notons Q l'application complètement positive déterminée par $Q(\lambda_g) = \phi(g)\lambda_g$ sur $vN(G)$.

Pour éviter les problèmes liés aux produits tensoriels infinis, et rester dans le même cadre que celui du chapitre 1, on va construire une dilatation de $(Q^n)_{n \leq N}$ pour un certain entier N. La construction d'une dilatation du semi-groupe Q^n tout entier est identique, mais avec un produit tensoriel infini.

Soit ρ une représentation unitaire de G sur un espace de Hilbert H et un état S sur H tel que $\phi(g) = tr(S\rho(g))$.

La représentation ρ étant donnée, on peut toujours trouver H et S par la construction GNS (cf [24]).

Soit $N \in \mathbb{N}$, on pose $\mathcal{W} = \mathcal{B}(H^{\otimes N})$, $\omega = S^{\otimes N}$. Pour tout $n \leq N$, $\mathcal{W}_n$ est l'algèbre des opérateurs de la forme $a \otimes Id$, avec $a \in \mathcal{B}(H^{\otimes n})$ sur la décomposition $H^{\otimes N} = H^{\otimes n} \otimes H^{\otimes (N-n)}$.

On définit des espérances conditionnelles $E_n : \mathcal{W} \to \mathcal{W}_n$ par

$$E_n(a_1 \otimes \ldots \otimes a_n \otimes a_{n+1} \otimes \ldots \otimes a_N)$$
$$= a_1 \otimes \ldots \otimes a_n \otimes tr(Sa_{n+1})Id \otimes \ldots \otimes tr(Sa_N)Id$$

On vérifie sans peine que les E_n vérifient $E_n E_m = E_{n \wedge m}$.

On construit des morphismes $j_n : vN(G) \to \mathcal{W}$ en posant $j_n(\lambda_g) = \rho(g)^{\otimes n} \otimes Id^{\otimes (N-n)}$

PROPOSITION 19 – $(\mathcal{W}, \omega, \mathcal{W}_n, E_n, j_n)_{n \leq N}$ est une dilatation de $(Q^n)_{n \leq N}$.

Démonstration

Il suffit de vérifier la propriété de Markov $E_n \circ j_{n+m} = j_n \circ Q^m$ sur un élément de la forme λ_g, or on a

$$E_n \circ j_{n+m}(\lambda_g) = \phi(g)^m \rho(g)^{\otimes n} \otimes Id^{\otimes (N-n)} = j_n \circ Q^m(\lambda_g)$$

PROPOSITION 20 – Soit $K \subset G$ un sous-groupe fermé, $vN_G(K) \subset vN(G)$ la sous algèbre de von Neumann engendrée par les éléments λ_h, $h \in K$ alors les restrictions des j_n à $vN_G(K)$ forment une dilatation de $Q_{|vN_G(K)}$. Si K est commutatif, les algèbres $j_n(vN_G(K))$ commutent.

Démonstration

La première partie est identique à celle de la proposition précédente.

Soient $m \leq n \in \mathbb{N}^*$, $h, h' \in H$ il est immédiat que $j_n(\lambda_h)$ commute avec $j_m(\lambda_{h'})$, d'où la seconde partie.

Dans le cas où G est un groupe commutatif, $vN(G)$ est isomorphe à $L^\infty(\hat{G})$ où $\hat{G}$ est le groupe dual de G, et l'état ν provient d'une mesure de probabilité sur $\hat{G}$ dont la transformée de Fourier est ϕ. La construction que nous avons faite ci-dessus se ramène alors à la construction usuelle d'une marche aléatoire sur $\hat{G}$ comme somme d'une famille de variables aléatoires indépendantes équidistribuées.

Dans le cas général, le théorème de Peter-Weyl (cf Bourbaki [14]) montre que l'espace $L^2(G)$ est une somme directe orthogonale $\oplus_{\chi \in \hat{G}} E_\chi$, où $\hat{G}$ est l'ensemble des classes d'équivalence de représentations irréductibles de G, et E_χ l'espace des fonctions coefficients de cette représentation. On en déduit que $vN(G) = \oplus_{\chi \in \hat{G}} M_\chi$ (la somme étant prise au sens des algèbres de von Neumann), où M_χ est l'algèbre de convolution à gauche par les fonctions de E_χ, qui est isomorphe à $M_d(\mathbb{C})$, d étant la dimension des représentations de classe χ.

Soit Z le centre de l'algèbre $vN(G)$, il résulte de la discussion ci-dessus que Z est isomorphe à l'algèbre $L^\infty(\hat{G})$.

PROPOSITION 21 – Les algèbres $j_n(Z)$ commutent. Si la fonction ϕ est centrale, alors Q envoie Z dans elle-même, et les $j_{n|Z}$ forment une dilatation de $Q^n_{|Z}$.

Démonstration

Soient $m \leq n \in \mathbb{N}^*$, $c \in Z$ et $g \in G$, alors $j_n(\lambda_g) = j_m(\lambda_g)\big(j_m(\lambda_g)^{-1}j_n(\lambda_g)\big)$ et $j_m(\lambda_g)^{-1}j_n(\lambda_g)$ commute avec $j_m(vN(G))$, or $j_m(c)$ commute avec $j_m(\lambda_g)$ car $c \in Z$ donc $j_m(c)$ commute avec $j_n(\lambda_g)$, on en déduit que $j_m(c)$ commute avec $j_n(vN(G))$ et donc que $j_m(Z)$ et $j_n(Z)$ commutent.

L'algèbre Z est engendrée en tant qu'algèbre de von Neumann par les opérateurs de la forme $\int_G k(g)\lambda_g dg$ où k est une fonction continue centrale. L'image par Q d'un tel opérateur est $\int_G \phi(g)k(g)\lambda_g dg$ qui est dans le centre Z. On en conclut par passage à la limite que Q préserve Z.

Pour montrer que les $j_{n|Z}$ forment une dilatation de $Q^n_{|Z}$ il suffit de montrer que pour tout m l' espérance conditionnelle E_m envoie l'algèbre engendrée par $j_{m+1}(Z) \ldots j_{m+k}(Z) \ldots$ sur celle engendrée par $j_1(Z), \ldots j_m(Z)$. Par récurrence il suffit de montrer que $E_m(j_{m+1}(Z)) \subset j_m(Z)$. Or, si $c = \int_G k(g)\lambda_g dg$ avec k centrale continue, il résulte de la définition de E_n que $E_n j_{m+1}(c) \in j_n(c)$, et par passage à la limite cela est vrai pour tout $c \in Z$.

L'algèbre de von Neumann engendrée par les $j_n(Z)$ est une algèbre commutative, donc il existe un espace de probabilité $(\Omega, \mathcal{F}, P)$ et des variables aléatoires $S_n : \Omega \to \hat{G}$, telles que $j_n(f) = f \circ S_n$. La restriction de Q à Z provient d'un noyau markovien N sur $\hat{G}$ et les variables S_n forment donc une chaîne de Markov sur $\hat{G}$ de noyau de transition N.

Nous allons retrouver la marche de Bernoulli quantique du chapitre 1, et le processus de Spin de la façon suivante. On prend $G = SU(2)$, $\phi(g) = \frac{1}{2}tr(g)$. Si ρ est une représentation de dimension finie de $SU(2)$ on peut la prolonger à son algèbre de Lie $su(2)$ en posant $\rho(\gamma) = \lim_{t\to 0} \frac{1}{t}\rho(\exp t\gamma - Id)$. Dans le cas de la fonction $\phi = \frac{1}{2}tr$, on peut prendre $H = \mathbb{C}^2$, ρ étant la représentation identique de dimension 2 de $SU(2)$, et $S = \frac{1}{2}Id$, dans la construction de la dilatation. On a alors $\mathcal{W} = \mathbb{C}^{2^N}$, et on vérifie que, avec les notations du chapitre 1, on a $j_n(i\sigma_x) = X_n$, $j_n(i\sigma_y) = Y_n$, et $j_n(i\sigma_z) = Z_n$. Les composantes de la marche de Bernoulli quantique sont donc obtenues en restreignant la dilatation j_n à des sous-algèbres engendrées par des sous-groupes à un paramètre de SU(2).

Pour ce qui est du processus de spin, nous allons faire intervenir le centre de l'algèbre de von Neumann. La théorie des représentation du groupe $SU(2)$ montre qu'il existe à isomorphisme près une unique représentation irréductible de dimension n pour tout entier $n \geq 1$. On a donc une identification naturelle $Z \sim L^\infty(\mathbb{N}^*)$. Les matrices de Pauli, en tant qu'éléments de l'algèbre de Lie complexifiée de $SU(2)$ définissent des dérivations invariantes à droite sur $L^2(SU(2))$ que nous notons X, Y et Z. L'opérateur auto-adjoint non-borné invariant à droite $\Sigma = \sqrt{X^2 + Y^2 + Z^2 + Id}$ sur $L^2(SU(2))$ commute avec tous les opérateurs λ_g. Il est diagonalisable et admet la valeur propre n sur le sous-espace de $L^2(SU(2))$ formé des fonctions coefficients de la représentation irréductible de dimension n de $SU(2)$. Cet opérateur s'interprète comme la fonction $n \mapsto n$ sur $\mathbb{N}^*$ quand on identifie $\hat{G}$ et $\mathbb{N}^*$. Remarquons que Σ n'est pas dans Z car il n'est pas borné.

On a $j_n(\Sigma) = \Sigma_n$, donc le processus de Spin est obtenu grâce à l'opérateur Σ.

Nous allons retrouver directement les probabilités de transition de cette chaîne de Markov en considérant la restriction de Q à $Z \sim L^\infty(\mathbb{N}^*)$.

Soit ρ_n une représentation irréductible de dimension n de $SU(2)$, son caractère est la fonction $\chi_n : g \mapsto tr(\rho_n(g))$ sur $SU(2)$. Cette fonction est entièrement déterminée par sa restiction aux matrices $g_\theta = \begin{pmatrix} e^{i\theta} & 0 \\ 0 & e^{-i\theta} \end{pmatrix}$ et on a $\chi_n(g_\theta) = \frac{\sin n\theta}{\sin \theta}$. De cette formule on déduit la formule de Clebsch-Gordon $\chi_n \chi_2 = \chi_{n+1} + \chi_{n-1}$. Remarquons que $\chi_2 = 2\phi$.

L'élément de Z qui correspond à l'indicatrice de n dans l'isomorphisme $Z \sim L^\infty(\mathbb{N}^*)$ est $n \int_{SU(2)} \chi_n(g) dg$. On en déduit que $Q(1_n) = \frac{1}{2}(\frac{n}{n+1} 1_{n+1} + \frac{n}{n-1} 1_{n-1})$ et que l'opérateur Q est bien l'opérateur markovien associé aux probabilités de transition du processus de Spin.

6.3 Equations différentielles linéaires

Nous reprenons les notations de la fin du chapitre 5. On a un espace de Fock $\Gamma(L^2(\mathbb{R}+) \otimes \mathbb{C}^d)$ sur lequel sont définis les opérateurs de création, annihilation et conservation notés collectivement a^ε. Le paramètre ε peut prendre les valeurs $+j$, $-j$, $0jk$, ou $\cdot$ avec $1 \le j, k \le d$. Rappelons que l'on a posé $da_t = dt$.

On a également un espace de Hilbert initial H.

Nous allons résoudre une équation différentielle stochastique non-commutative linéaire à coefficients constants.

On se donne des opérateurs bornés L^ε sur H, et on cherche à résoudre l'équation

$$dV_t = V_t \left(\sum_\varepsilon L^\varepsilon da_t^\varepsilon \right) \tag{99}$$

Ici les opérateurs L^ε sont étendus à $H \otimes \Gamma(L^2(\mathbb{R}+) \otimes \mathbb{C}^d)$ par $L^\varepsilon \otimes Id$. Cette équation doit être interprétée comme une équation intégrale, c'est à dire que l'on doit avoir d'une part $V_t L^\varepsilon$ défini sur $H \otimes \Xi(S)$, intégrable, et d'autre part,

$$V_t = V_0 + \sum_\varepsilon \int_0^t V_s L^\varepsilon da_s^\varepsilon$$

pour tout $t \ge 0$.

Soit $S = L^2 \cap L^\infty(\mathbb{R}_+)$.

THÉORÈME 8 – Soit V_0 un opérateur borné sur H, il existe un unique processus V_t d'opérateurs de domaine $H \otimes \Xi(S)$ tel que

$i)$ $\qquad \forall t \ge 0 \quad V_t = V_0 + \sum_\varepsilon \int_0^t V_s L^\varepsilon da_s^\varepsilon$

$ii)$ $\qquad \forall t \ge 0 \; \forall u \in S \quad \sup_{|a| \le 1, s \le t} |V_s(a \otimes \xi(u))| < +\infty$

(V_0 est identifié avec l'opérateur $V_0 \otimes Id$ dans i))

Démonstration

Commençons par l'existence. On va utiliser la méthode d'itération de Picard. Posons

$$\forall t \geq 0 \quad V_t^{(0)} = V_0$$

$$V_t^{(n+1)} = V_0 + \sum_\varepsilon \int_0^t V_s^{(n)} L^\varepsilon da_s^\varepsilon$$

On vérifie par récurrence sur n en utilisant la remarque (1) après le théorème 7 que le processus $V^{(n)}$ est bien défini sur le domaine $H \otimes \Xi(S)$ et que d'après l'inégalité (86)

$$\sup_{|a| \leq 1, s \leq t} |V_s^{(n)} a \otimes \xi(u)|^2 < +\infty$$

D'autre part on a

$$(V_t^{(n+1)} - V_t^{(n)})a \otimes \xi(u) = \sum_\varepsilon \left(\int_0^t (V_s^{(n)} - V_s^{(n-1)}) L^\varepsilon da_s^\varepsilon \right) a \otimes \xi(u)$$

d'où, d'après l'inégalité (86)

$$|(V_t^{(n+1)} - V_t^{(n)})a \otimes \xi(u)|^2 \leq K(u) \sum_\varepsilon \int_0^t |(V_s^{(n)} - V_s^{(n-1)}) L^\varepsilon a \otimes \xi(u)|^2 ds$$

où $K(u)$ est la constante $(|u| + \sqrt{1 + |u|^2})(1 + |u|_\infty^2)|\xi(u)|^2$

Si on pose $\alpha_n(t) = \sup_{|a| \leq 1, s \leq t} |(V_s^{(n+1)} - V_s^{(n)})a \otimes \xi(u)|^2$ alors cette inégalité montre que

$$\alpha_n(t) \leq K' \int_0^t \alpha_{n-1}(s) ds$$

avec $K' = K(u)\left(\sum_\varepsilon \|L^\varepsilon\|^2 \right)$, d'où l'on déduit par récurrence sur n que

$$\alpha_n(t) \leq \frac{K'^n}{n!} t^n |V_0 a \otimes \xi(u)|^2$$

La série $\sum_{n=0}^\infty V_t^{(n+1)} - V_t^{(n)}$ converge donc fortement sur le domaine $H \otimes \Xi(S)$. Soit V_t la somme de cette série.

On a $\sup_{|a| \leq 1, s \leq t} |V_t a \otimes \xi(u)|^2 < +\infty$ et $\int_0^t |(V_s - V_s^{(n)})a \otimes \xi(u)|^2 ds \to_{n \to +\infty} 0$ ce qui entraine

$$\left(\sum_\varepsilon \int_0^t V_s^{(n)} L^\varepsilon da_s^\varepsilon \right) a \otimes \xi(u) \to_{n \to +\infty} \left(\sum_\varepsilon \int_0^t V_s L^\varepsilon da_s^\varepsilon \right) a \otimes \xi(u)$$

et donc V est solution de l'équation (99).

Passons maintenant à l'unicité. Soit V' une autre solution, on pose $Y_t = V_t - V'_t$. L'inégalité (86) donne encore

$$|Y_t a \otimes \xi(u)|^2 \leq K(u) \sum_\varepsilon \int_0^t |Y_s L^\varepsilon a \otimes \xi(u)|^2 ds$$

En posant $\beta(t) = \sup_{|a| \leq 1 \ s \leq t} |Y_s a \otimes \xi(u)|^2$ on voit que $0 \leq \beta(t) \leq K' \int_0^t \beta(s) ds$ ce qui entraine que $\beta \equiv 0$.

Nous allons nous intéresser aux conditions sur les coefficients L^ε qui font que les opérateurs V_t peuvent se prolonger en des opérateurs isométriques. Afin de ne pas obscurcir la discussion par des considérations d'indices, on va traiter seulement le cas $d = 1$, le paramètre ε peut donc prendre les valeurs $-, 0, +$ ou $\cdot$. Je laisse au lecteur le soin d'énoncer les résultats analogues pour $d \geq 2$ (cf [51] et [60]).

Commençons par chercher des conditions nécessaires pour que la solution V_t de (99) soit isométrique, avec une condition initiale $V_0 = Id$.

Soient $x = a \otimes \xi(u)$ et $y = b \otimes \xi(v) \in H \otimes \Xi(S)$. Pour simplifier les notations on pose $dM_s = \sum_\varepsilon L^\varepsilon da_s^\varepsilon$.

Calculons, à l'aide de la formule d'Itô

$$< V_t x, V_t y > = < x, y > + < x, \int_0^t V_s dM_s y > + < \int_0^t V_s dM_s x, y >$$

$$+ < \int_0^t V_s dM_s x, \int_0^t V_s dM_s y > ds$$

$$= < x, y > + \int_0^t < V_s R_s x, y > ds + \int_0^t < x, V_s R_s^* y > ds$$

$$+ \int_0^t < V_s R_s x, (V_s y - y) > ds + \int_0^t < (V_s x - x), V_s R_s^* y > ds$$

$$+ \int_0^t < V_s T_s^1 x, V_s T_s^2 y > ds$$

$$= < x, y > + \int_0^t < V_s R_s x, V_s y > ds + \int_0^t < V_s x, V_s R_s^* y >$$

$$+ \int_0^t < V_s T_s^1 x, V_s T_s^2 y > ds$$

où on a posé

$$R_s = u(s)L^- + u\bar{v}(s)L^0 + \bar{v}(s)L^+ + L^\cdot$$

$$T_s^1 = L^+ + u(s)L^0 \qquad T_s^2 = L^+ + v(s)L^0$$

L'hypothèse que V est isométrique implique donc

$$\forall t \geq 0 \ \forall u, v \in S \ \forall a, b \in H \quad \int_0^t < R_s x, y > + < x, R_s^* y > + < T_s^1 x, T_s^2 y > ds = 0$$

ce qui entraine que $\forall a, b \in H$

$$< L^\cdot a, b > + < a, L^\cdot b > + < L^+ a, L^+ b > = 0 \tag{100}$$

$$< L^+a, b > + < a, L^-b > + < L^+a, L^0b >= 0 \qquad (101)$$

$$< L^-a, b > + < a, L^+b > + < L^0a, L^+b >= 0 \qquad (102)$$

$$< L^0a, b > + < a, L^0b > + < L^0a, L^0b >= 0 \qquad (103)$$

Les conditions (101) et (102) sont équivalentes, et on peut mettre (100), (101), (102), (103) sous la forme équivalente

$$L^0 + (L^0)^* + (L^0)^* L^0 = 0 \qquad (104)$$

$$L^+ + (L^-)^* + (L^0)^* L^+ = 0 \qquad (105)$$

$$L^{\cdot} + (L^{\cdot})^* + (L^+)^* L^+ = 0 \qquad (106)$$

Ces conditions sont nécessaires. Montrons qu'elles sont suffisantes pour que V_t soit isométrique. Si ces conditions sont vérifiées, on obtient en appliquant la formule d'Itô

$$< V_t x, V_t y > - < x, y > = \int_0^t \big(< V_s R_s x, V_s y > - < R_s x, y > \big) ds$$
$$+ \int_0^t \big(< V_s x, V_s R_s^* y > \quad < x, R_s^* y > \big) ds$$
$$+ \int_0^t \big(< V_s T_s^1 x, V_s T_s^2 y > - < T_s^1 x, T_s^2 y > \big) ds$$

Posons $\psi(t) = \sup_{|a|,|b| \le 1 \; s \le t} | < V_s x, V_s y > - < x, y > |$, il existe une constante K telle que $0 \le \psi(t) \le K \int_0^t \psi(s) ds$ ce qui entraine que $\psi \equiv 0$.

Les conditions (104), (105), (106) sont donc nécessaires et suffisantes pour que V soit prolongeable en un processus d'isométries.

Soit L^ε une famille d'opérateurs bornés satisfaisant les conditions (104), (105), et (106), on voit facilement que ces conditions sont équivalentes à l'existence d'un opérateur isométrique W, un opérateur borné J, et un opérateur auto-adjoint borné H tels que

$$L^0 = W - I$$
$$L^+ = J^*$$
$$L^- = -JW$$
$$L^{\cdot} = iH - \frac{1}{2} J J^*$$

L'équation différentielle (99) s'écrit alors

$$dV_t = V_t \big((W - I) da_t^0 - JW da_t^- + J^* da_t^+ + (iH - \frac{1}{2} J J^*) dt \big) \qquad (107)$$

V étant isométrique il possède un adjoint borné.

Les formules $< V_t^* x, y >= < x, V_t y >$ et (77) montrent que le processus V^* est solution de l'équation différentielle stochastique non-commutative

$$dV_t^* = \big((W^* - I) da_t^0 + J da_t^- - W^* J^* da_t^+ + (-iH - \frac{1}{2} J J^* dt) \big) V_t^* \qquad (108)$$

(remarquons qu'il ne s'agit *pas* d'une équation du même type que (99), car les opérateurs L^ε ne commutent pas a priori avec V.)

Supposons maintenant que l'opérateur W soit non seulement isométrique, mais aussi unitaire. En appliquant la formule d'Itô à l'expression

$$< V_t^* a \otimes \xi(u), V_t^* b \otimes \xi(v) >$$

à l'aide de l'équation (108) on constate que l'opérateur V_t^* est isométrique, et donc V_t est unitaire.

6.4 Dilatations de semi-groupes complètement positifs

On considère un opérateur unitaire W, un opérateur borné J, et un opérateur auto-adjoint borné H, et on note U le processus d'opérateurs unitaires solution de l'équation (107). Le processus adjoint U_t^* vérifie donc l'équation (108).

Nous allons utiliser les processus d'opérateurs U_t et U_t^* pour construire une dilatation d'un semi-groupe complètement positif sur $\mathcal{B}(H)$. Commençons par mettre en place les objets nécessaires.

Soit $\mathcal{W} = \mathcal{B}(H \otimes \Gamma(L^2_{\mathbf{C}}(\mathbf{R}_+)))$. On a une filtration naturelle en prenant pour $\mathcal{W}_t$ la sous-algèbre des opérateurs bornés de la forme $A \otimes Id$ sur l'espace $(H \otimes \Gamma_{t]}) \otimes \Gamma_{[t}$, avec $A \in \mathcal{B}(H \otimes \Gamma_{t]})$.

Donnons nous un état quelconque S sur $\mathcal{B}(H)$, et définissons l'état ω sur $\mathcal{W}$ par $\omega(A \otimes X) = tr(SA) < X\Omega, \Omega >$ pour $A \in \mathcal{B}(H)$ et $X \in \mathcal{B}(\Gamma)$.

Soit $\mathbf{E}_t$ le projecteur orthogonal sur le sous-espace $H \otimes \Gamma_{t]} \otimes \Omega_{[t}$.

Soient $a \otimes m_{t]} \otimes m'_{[t} \in H \otimes \Gamma_{t]} \otimes \Gamma_{[t}$, et $X \in \mathcal{W}$ alors

$$\mathbf{E}_t \circ X(a \otimes m_{t]} \otimes m'_{[t}) = n_{t]} \otimes \Omega_{[t}$$

avec $n_{[t} \in H \otimes \Gamma_{t]}$, et on pose

$$(E_t X)(a \otimes m_{t]} \otimes m'_{[t}) = n_{t]} \otimes m'_{[t}$$

On définit ainsi un opérateur $E_t X \in \mathcal{W}_t$, et les applications $E_t : \mathcal{W} \to \mathcal{W}_t$, forment une famille d'espérances conditionnelles qui vérifient la condition $E_t E_s = E_{s \wedge t}$.

Soit $x \in \mathcal{B}(H)$, on pose

$$A^{\cdot} x = i[H, x] - \frac{1}{2}(JJ^* x + xJJ^* - 2JWxW^* J^*)$$

L'application $A^{\cdot}$ est linéaire et bornée sur l'espace de Banach $\mathcal{B}(H)$.

On définit une famille de morphismes $j_t : \mathcal{B}(H) \to \mathcal{W}_t$, en posant

$$j_t(x) = U_t \circ (x \otimes Id) \circ U_t^*$$

Les U_t étant unitaires, il est clair que j_t est un morphisme à valeurs dans $\mathcal{W}$, qui préserve l'identité, et comme U est un processus adapté, il est à valeurs dans $\mathcal{W}_t$.

Nous pouvons maintenant énoncer le principal résultat de ce paragraphe.

THÉORÈME 9 – *Les applications $T_t = e^{tA}$ forment un semi-groupe d'applications complètement positives sur $\mathcal{B}(H)$, et $(\mathcal{W}, \omega, \mathcal{W}_t, E_t, j_t)_{t \in \mathbf{R}_+}$ est une dilatation de ce semi-groupe.*

Avant de démontrer ce théorème, nous allons établir deux résultats importants.
On définit des opérateurs bornés A^ε, $\varepsilon = -, 0, +$ sur $\mathcal{B}(H)$ par
$A^0 x = WxW^* - x$, $A^+ x = J^* x - WxW^* J^*$, $A^- x = xJ - JWxW^*$.

PROPOSITION 22 – Pour tout $t \geq 0$ et tout $x \in \mathcal{B}(H)$, on a

$$j_t(x) = x \otimes Id + \sum_\varepsilon \int_0^t j_t(A^\varepsilon x) da_t^\varepsilon$$

Démonstration
Il suffit de montrer que pour tous $a, b \in H$, tous $u, v \in S$ on a

$$< j_t(x)a \otimes \xi(u), b \otimes \xi(v) > = < (x \otimes Id + \sum_\varepsilon \int_0^t j_t(A^\varepsilon x) da_t^\varepsilon) a \otimes \xi(u), b \otimes \xi(v) >$$

or

$$< j_t(x)a \otimes \xi(u), b \otimes \xi(v) > = < (x \otimes Id) U_t^* a \otimes \xi(u), U_t^* b \otimes \xi(v) >$$

et un calcul long, mais sans difficulté utilisant la formule d'Itô et l'équation (108) vérifiée par U^* permet de montrer la proposition.

PROPOSITION 23 – Soit X_t un processus d'opérateurs bornés tel qu'il existe des processus adaptés K_t^ε d'opérateurs bornés vérifiant $X_t = X_0 + \sum_\varepsilon \int_0^t K_s^\varepsilon da_s^\varepsilon$ pour tout $t \leq 0$, alors $\forall s, t \in \mathbf{R}_+$ $E_t X_{t+s} = X_t + \int_t^{t+s} E_t K_r \, dr$.

Démonstration
Soient $a, b \in H$ et $u, v \in S$, par définition de E_t on a

$$< E_t X_{t+s} a \otimes \xi(u), b \otimes \xi(v) > = < X_{t+s} a \otimes \xi(u_{t]}), b \otimes \xi(v_{t]}) > < \xi(u_{[t}), \xi(v_{[t}) >$$

$$= \sum_{\varepsilon \neq \cdot} \int_0^t < K_r^\varepsilon a \otimes \xi(u_{t]}), b \otimes \xi(v_{t]}) > < \xi(u_{[t}), \xi(v_{[t}) > \kappa^\varepsilon(r) dr$$

$$+ \int_0^{t+s} < K_r a \otimes \xi(u_{t]}), b \otimes \xi(v_{t]}) > < \xi(u_{[t}), \xi(v_{[t}) > dr$$

$$= \sum_{\varepsilon \neq \cdot} \int_0^t < K_r^\varepsilon a \otimes \xi(u), b \otimes \xi(v) > \kappa^\varepsilon(u) dr$$

$$+ \int_0^{t+s} < E_t K_r a \otimes \xi(u), b \otimes \xi(v) > dr$$

$$= < X_t a \otimes \xi(u), b \otimes \xi(v) >$$

$$+ \int_t^{t+s} < E_t K_r a \otimes \xi(u), b \otimes \xi(v) > dr$$

on en déduit la proposition.

Nous pouvons maintenant passer à la démonstration du théorème (9).
D'après les propositions (22) et (23) on a pour tous $s, t \in \mathbf{R}_+$ et $x \in \mathcal{B}(H)$

$$E_t j_{t+s}(x) = j_t(x) + \int_t^{t+s} E_t j_r(A^\cdot x) dr$$

On en déduit que $\frac{d}{ds}\big(E_t j_{t+s}(x)\big) = E_t j_{t+s}(A\,x)$. En résolvant cette équation différentielle, on voit que $E_t j_{t+s}(x) = j_t(e^{sA}\,x)$, d'où la propriété de Markov de j_t.

En particulier, $T_t(x) = E_0 \circ j_t(x) \circ E_0$. T_t est donc la composée de deux applications dont on voit facilement qu'elles sont complètement positives, d'une part $x \mapsto j_t(x)$ et d'autre part $X \mapsto E_0 X E_0$ de $\mathcal{W}$ dans $\mathcal{B}(H)$, donc T_t est un semi-groupe d'applications complètement positives sur $\mathcal{B}(H)$ et $(\mathcal{W}, \omega, \mathcal{W}_t, j_t, E_t)$ en est une dilatation.

6.5 Application aux processus de Markov

Les équations différentielles stochastiques ordinaires donnent un moyen de construire des diffusions, c'est à dire des processus de Markov à valeurs dans des variétés, dont le générateur est donné par un opérateur différentiel d'ordre 2 (voir [43]). Quitte à se placer dans une carte locale, une telle équation a la forme suivante

$$dX_t = \sigma(X_t)dB_t + b(X_t)dt \tag{109}$$

C'est une équation différentielle stochastique vectorielle dans laquelle le processus inconnu X est à valeurs dans un ouvert O de $\mathbb{R}^n$, σ est une fonction sur O à valeurs dans les matrices réelles $n \times d$, b une fonction sur O à valeurs dans $\mathbb{R}^n$, et B_t un mouvement brownien $d-$dimensionnel. En utilisant la formule d'Itô on peut écrire pour toute fonction f de classe C^2 sur la variété

$$df(X_t) = Lf(X_t)dB_t + Af(X_t)dt \tag{110}$$

Ici, $L = (L^1, \ldots, L^d)$ est une famille de d champs de vecteurs sur la variété, et A est un opérateur différentiel d'ordre 2, le générateur de la diffusion. Dans une carte locale, les champs de vecteurs L et l'opérateur A s'expriment au moyen de σ et b (cf [43]).

L'équation (110) peut être considérée comme une formulation intrinsèque de l'équation en coordonnées locales (109).

L'application qui à une fonction f sur la variété fait correspondre la variable aléatoire $f \circ X_t$, considérée comme opérateur de multiplication sur l'espace de Fock du mouvement brownien, est un morphisme d'algèbres. Désignons le par k_t. Cela permet d'écrire (109) sous la forme

$$dk_t(f) = k_t(A^+ f)da_t^+ + k_t(A^- f)da_t^- + k_t(A\,f)dt$$

avec $A^+ = A^- = L$ et $A\, = A$, réminiscente de la proposition 22 (mais ici on est sur un espace de Fock de multiplicité $d \geq 1$).

Maintenant que nous avons mis en évidence la similitude entre les calculs du paragraphe 6.4 et la construction des processus de diffusion, nous allons utiliser ces idées pour construire des processus de Markov discontinus en résolvant des équations différentielles du type (107). L'idée de base est la suivante, partant du processus de Markov X_t, on exhibe une martingale normale d-dimensionnelle, c'est à dire une famille de martingales M^j, $j = 1, \ldots, d$ telles que $< M^j, M^k >_t = \delta_{j,k} t$ qui permet de reconstruire le processus de Markov X_t par une équation du type (110) dans laquelle le mouvement brownien B est remplacé par la martingale normale M. L'inconvénient de la formulation de cette équation est que la martingale M n'est pas canonique. Nous verrons que si on interprète cette martingale comme une famille d'opérateurs de multiplication sur l'espace de Fock, ce qui est possible si elle possède la propriété de représentation chaotique, en utilisant le lemme 18 pour les exprimer comme des intégrales stochastiques non-commutatives d'opérateurs par rapport aux processus de base da^ε, on obtient une équation non-commutative du type de la proposition 24, que l'on résout grâce aux résultats du paragraphe précédent.

Nous allons traiter deux exemples simples. Commençons par le processus de naissance pur.

Il s'agit d'un processus de Markov sur $\mathbb{Z}$ de générateur

$$Af(x) = a_x\big(f(x+1) - f(x)\big)$$

Nous supposerons que les coefficients a_x vérifient une inégalité $0 < \frac{1}{c} < a_x < c$.
Le semi-groupe de générateur A est $T_t = e^{tA}$ sur $L^\infty(\mathbb{Z})$, et on peut décrire les trajectoires du processus correspondant de la façon suivante. Le processus part de X_0 où il reste pendant un temps de loi exponentielle de paramètre a_{X_0} puis il saute en $X_0 + 1$ où il reste un temps indépendant du premier temps de saut, de loi exponentielle de paramètre a_{X_0+1} avant de sauter en $X_0 + 2$, et ainsi de suite.
Supposons X_0 déterministe. Nous allons tout d'abord identifier l'espace L^2 du processus avec un espace de Fock en utilisant une martingale normale qui possède la propriété de représentation chaotique.
Cette martingale est

$$M_t = \sum_{s \leq t} 1_{\{X_{s-} \neq X_s\}} \frac{1}{\sqrt{a_{X_{s-}}}} - \int_0^t \sqrt{a_{X_{s-}}}\,ds$$

Elle est à variation finie, et le lemme facile suivant est laissé au lecteur.

LEMME – *M est une martingale de carré intégrable de crochets*

$$< M, M >_t = t$$

et

$$[M, M]_t = t + \int_0^t \frac{1}{\sqrt{a_{X_{s-}}}}\,dM_s$$

La martingale M est adaptée à la filtration du processus X, et réciproquement, le processus X peut être reconstruit à partir de M et de X_0 en résolvant l'équation

$$df(X_t) = \sqrt{a_{X_{t-}}}\big(f(X_{t-}+1) - f(X_{t-})\big)dM_t + a_{X_{t-}}\big(f(X_{t-}+1) - f(X_{t-})\big)dt$$

pour toute fonction f sur $\mathbb{Z}$.
Si on pose $Tf(x) = \sqrt{a_x}\big(f(x+1) - f(x)\big)$ on peut l'écrire

$$df(X_t) = Tf(X_{t-})dM_t + Af(X_{t-})dt$$

LEMME – *La martingale M possède la propriété de représentation chaotique.*

Ce lemme découle d'un résultat plus général dû à Emery ([27] Théorème 5).

Soit $\mathcal{X}$ la tribu du processus X, le lemme précédent entraine que $L^2(\mathcal{X}) \sim \Gamma(L^2_{\mathbb{C}}(\mathbb{R}_+))$. En appliquant la proposition 18 à M on obtient

PROPOSITION 24 – L'opérateur de multiplication par M_t en tant qu'opérateur sur l'espace de Fock s'écrit

$$M_t = \int_0^t \frac{1}{\sqrt{a_{X_{s-}}}} da_s^0 + a_t^+ + a_t^-$$

Soit $H = L^2(\mathbb{Z})$ ($\mathbb{Z}$ étant muni de la mesure de comptage), l'algèbre $L^\infty(\mathbb{Z})$ agissant par multiplication sur $L^2(\mathbb{Z})$ est une sous-algèbre de $\mathcal{B}(H)$. En utilisant l'expression précédente pour l'opérateur de multiplication par M, nous allons étudier l'équation différentielle suivante

$$dk_t(x) = k_t(A^0 x) da_t^0 + k_t(A^+ x) da_t^+ + k_t(A^- x) da_t^- + k_t(A^. x) dt \tag{111}$$

où k_t est un morphisme de $L^\infty(\mathbb{Z})$ dans $\mathcal{W}_t$ (avec les notations de 6.3), et les applications $A^\varepsilon : L^\infty(\mathbb{Z}) \to L^\infty(\mathbb{Z})$ sont données par $A^0(f)(n) = f(n+1) - f(n)$, $A^+ = T = A^-$, et $A^. = A$.
Soit $W : L^2(\mathbb{Z}) \to L^2(\mathbb{Z})$ l'opérateur unitaire $Wf(n) = f(n+1)$, et J l'opérateur borné $Jf(n) = \sqrt{a_n} f(n)$, alors en tant qu'opérateurs sur $L^2(\mathbb{Z})$ on a pour tout $x \in L^\infty(\mathbb{Z})$
$A^0 x = WxW^* - x$, $A^+ x = J^* x - WxW^* J^*$, $A^- x = xJ - JWxW^*$
et $A^. x = -\frac{1}{2}(JJ^* x + xJJ^* - 2JWxW^* J^*)$.
Les formules ci-dessus définissent en fait des applications sur $\mathcal{B}(H)$ tout entier. On peut maintenant utiliser les résultat du paragraphe précédent. Soit U_t la solution unitaire de l'équation

$$dU_t = U_t\left((W - I) da_t^0 - JW da_t^- + J^* da_t^+ - \frac{1}{2} JJ^* dt\right)$$

avec $U_0 = Id$.
Avec les notations du paragraphe 6.3, $(\mathcal{W}, \omega, \mathcal{W}_t, j_t, E_t)$ est une dilatation du semi-groupe complètement positif $T_t = e^{tA}$, T_t préserve l'algèbre $L^\infty(\mathbb{Z})$ et sa restriction à cette algèbre est le semi-groupe markovien de générateur A. Les restrictions k_t des morphismes j_t à $L^\infty(\mathbb{Z})$ sont solution de (111)

PROPOSITION 25 – Les algèbres $j_t(L^\infty(\mathbb{Z}))$, $t \in \mathbb{R}_+$ commutent.

Démonstration
Soient $x, y \in L^\infty(\mathbb{Z})$, pour tout $t \geq 0$, $j_t(x)$ et $j_t(y)$ commutent.
Soit $s \geq 0$, on a $j_{t+s}(x) - j_t(x) = \sum_\varepsilon \int_t^{t+s} j_u(A^\varepsilon x) da_u^\varepsilon$, et on vérifie en utilisant (77) que

$$(j_{t+s}(x) - j_t(x)) j_t(y) = \sum_\varepsilon \int_t^{t+s} j_u(A^\varepsilon x) j_t(y) da_u^\varepsilon$$

et

$$j_t(y)(j_{t+s}(x) - j_t(x)) = \sum_\varepsilon \int_t^{t+s} j_t(y) j_u(A^\varepsilon x) da_u^\varepsilon$$

Posons

$$\alpha(r) = \sup_{\|x\|,\|y\|\leq 1 \ |a|\leq 1, \ s\leq r} |(j_t(y)j_{t+s}(x) - j_{t+s}(x)j_t(y))a \otimes \xi(u)|$$

D'après (86) il existe une constante K telle que $0 \leq \alpha(r) \leq K \int_0^r \alpha(s)ds$, d'où $\alpha \equiv 0$. On en déduit que $j_t(L^\infty(\mathbb{Z}))$ et $j_{t+s}(L^\infty(\mathbb{Z}))$ commutent.

Considérons l'algèbre de von Neumann commutative $\mathcal{A}$ engendrée par les algèbres $j_t(L^\infty(\mathbb{Z}))$. Il existe un espace de probabilité $(\Omega, \mathcal{F}, P)$ et des variables aléatoires $X_t : \Omega \to \mathbb{Z}$ telles que $(\mathcal{A}, \omega) \sim (L^\infty(\Omega, \mathcal{F}, P), P)$, $j_t(f) = f \circ X_t$ pour $f \in L^\infty(\mathbb{Z})$.

PROPOSITION 26 – *Le processus X_t est un processus de Markov sur $\mathbb{Z}$ de semi-groupe T_t.*

Démonstration
Soient $t_1 < \ldots < t_n \in \mathbb{R}_+$ et $f_1, \ldots, f_n \in L^\infty(\mathbb{Z})$, on a

$$E[f_1(X_{t_1}) \ldots f_n(X_{t_n})] = \omega(j_{t_1}(f_1) \ldots j_{t_n}(f_n))$$
$$= \omega(j_{t_1}(f_1) \ldots j_{t_{n-1}}(f_{n-1}T_{t_n-t_{n-1}}(f_n)))$$

car j_t est une dilatation de T_t.
On en déduit par récurrence sur n que l'on a bien

$$E[f_1(X_{t_1}) \ldots f_n(X_{t_n})] = E[f_1 T_{t_2-t_1}(f_2 \ldots T_{t_n-t_{n-1}}(f_n) \ldots)(X_{t_1})]$$

ce qui permet de conclure facilement.

Passons maintenant à un second exemple plus intéressant d'un point de vue théorique, celui des chaînes de Markov à espace d'états fini. Je vais donner rapidement la méthode, les détails sont semblables à ceux du processus de naissance.

Soit X une chaîne de Markov en temps continu sur un espace d'états fini E. Cette chaîne a pour générateur l'opérateur $Af(x) = \sum_{y \in E} p(x,y)(f(y) - f(x))$ où les coefficients $p(x,y)$ sont ≥ 0. Nous allons identifier E avec un groupe (pour fixer les idées, avec $\mathbb{Z}/d$ si E a d éléments), et supposer que $p(x,y) > 0$ pour $x \neq y$. On définit alors $d - 1$ martingales $(M^j)_{j \in E \setminus \{0\}}$ par

$$M_t^j = \sum_{s<t} \frac{1}{\sqrt{p(X_{s-}, X_s)}} 1_{\{X_s - X_{s-} = j\}} - \int_0^t \sqrt{p(X_s, X_s + j)}ds$$

Ces martingales sont normales, elles ont pour crochets

$$< M^j, M^k >_t = \delta_{jk}t$$

$$[M^j, M^k]_t = \delta_{jk}\left(t + \int_0^t \sqrt{p(X_{s-}, X_{s-} + j)}dM_s^j\right)$$

et de plus elles possèdent la propriété de représentation chaotique (cf [10]).
On en déduit que les opérateurs de multiplication sur l'espace de Fock correspondant sont donnés par les formules

$$M_t^j = \int_0^t \frac{1}{\sqrt{p(X_{s-}, X_{s-} + j)}} da_s^{0jj} + a_t^{+j} + a_t^{-j}$$

Le processus X peut être reconstruit à partir des martingales M^j en résolvant l'équation différentielle stochastique

$$df(X_t) = \sum_j \sqrt{p(X_{t-}, X_{t-} + j)}(f(X_{t-} + j) - f(X_{t-}))dM_t^j + Af(X_t)dt$$

pour toute fonction f sur E. En reportant l'expression des martingales M_t^j sur l'espace de Fock, on se ramène à résoudre l'équation

$$df(X_t) = Af(X_t)dt + \sum_{j \neq 0}(f(X_{t-} + j) - f(X_{t-}))da_t^{0jj}$$
$$+ \sqrt{p(X_{t-}, X_{t-} + j)}(f(X_{t-} + j) - f(X_{t-}))(da_t^{+j} + da_t^{-j})$$

Considérons $H = L^2(E)$ (avec la mesure de comptage sur E) et $L^\infty(E) \subset \mathcal{B}(H)$. On définit des opérateurs unitairess W^j, et des opérateurs bornés J^j, $j \in E\backslash\{0\}$ sur H par

$$W^j f(x) = f(x + j) \text{ et } J^j f(x) = \sqrt{p(x, x + j)}f(x)$$

L'équation

$$dU_t = U_t \sum_{j \neq 0}\left(W^d a_t^{0jj} + (J^j)^* da_t^{+j} - W^j J^j da_t^{-j} - \frac{1}{2}J^j(J^j)^* dt\right)$$

admet une solution composée d'opérateurs unitaire $(U_t)_{t \in \mathbf{R}_+}$.
On définit des morphismes $j_t : \mathcal{B}(H) \to \mathcal{W}$ par $j_t(x) = U_t \circ (x \otimes Id) \circ U_t^*$. On vérifie grâce à la formule d'Itô (multidimensionnelle cette fois-ci) que ces morphismes vérifient l'équation

$$dj_t(x) = j_t(A^\cdot x)dt + \sum_{j \neq 0}(j_t(A^{0jj}x)da_t^{0jj} + j_t(A^{+j}x)da_t^{+j} + j_t(A^jx)da_t^{-j})$$

avec $A^{0j}x = W^j x(W^j)^* - x$, $A^{+j}x = (J^j)^*x - W^j x(W^j)^*(J^j)^*$,
$A^{-j}x = xJ^j - J^j W^j x(W^j)^*$, et
$A^\cdot x = \sum_{j \neq 0} -\frac{1}{2}\left(J^j(J^j)^*x + xJ^j(J^j)^* - 2J^j W^j x(W^j)^*(J^j)^*\right)$.
On choisit un état S sur $\mathcal{B}(H)$ et on construit l'état ω sur $\mathcal{W}$ comme dans l'exemple précédent. On obtient alors que $(\mathcal{W}, \omega, W, E_t, j_t)$ est une dilatation du semi-groupe $T_t = e^{tA}$. Il est facile de voir que ce semi-groupe laisse l'algèbre $L^\infty(E) \subset \mathcal{B}(H)$ invariante, et que sa restriction est le semi-groupe de générateur A. On montre encore que les $j_t(L^\infty(E))$ commutent et on en déduit qu'il existe un espace $(\Omega, \mathcal{F}, P)$, et des variables aléatoires $X_t : \Omega \to \mathbb{Z}$ tels que l'algèbre de von Neumann engendrée par les $j_t(L^\infty(E))$ soit isomorphe à $L^\infty(\Omega, \mathcal{F}, P)$, la restriction de ω à cette algèbre soit l'espérance pour P, et pour tout $f \in L^\infty(E)$, on ait $j_t(f) = f \circ X_t$. Les variables X_t forment un processus de Markov de générateur A et on a donc réussi à construire la chaîne de Markov de générateur A en résolvant une équation de la forme (99).

Commentaires et bibliographie

Pour écrire ce cours, j'ai utilisé abondamment le livre de K.R.Parthasarathy [60], ainsi que le gros article de P.A.Meyer [50], qui a été repris et considérablement modifié dans le livre [51]. J'espére que le présent cours pourra servir d'introduction à la lecture de ces sources qui couvrent de nombreux sujets que je n'ai pas pu traiter ici.

Les probabilités quantiques sont un vaste domaine, les commentaires et les références bibliographiques qui suivent en reflètent seulement ma connaissance partielle, et n'ont pas la prétention d'être exhaustifs.

Chapitre 1 − L'addition de variables de Bernoulli quantiques a été considérée par Meyer dans [50] comme donnant une approximation discrète de l'espace de Fock (le "bébé Fock"), d'après une idée de J.L.Journé.

La loi du processus de spin a été calculée dans [11], et [86], où on pourra trouver la solution de l'exercice.

Le théorème limite 2 est une version élémentaire d'un résultat de L.Accardi et A.Bach (non publié) exposé dans [52].

Chapitre 2 − Le contenu de ce chapitre est tout à fait classique, et peut se trouver dans tout bon ouvrage sur la théorie spectrale. La proposition 3 est inspirée de [20].

Chapitre 3 − Ce chapitre est inspiré du chapitre III de [50].

Les opérateurs de création et d'annihilation ont été introduits sous leur forme matricielle par W.Heisenberg. Le modèle des relations d'Heisenberg étudié dans ce chapitre n'est pas le plus couramment utilisé en physique, où l'on préfère utiliser les opérateurs x et $i\frac{d}{dx}$ sur $L^2(\mathbb{R}, dx)$ (dx étant la mesure de Lebesgue) à la place de Q et P. L'opérateur $i\frac{d}{dx}$ sert à calculer le moment d'une particule, et x sa position. Dans le paragraphe 3.8 pour tout élément de $SL(2, \mathbb{R})$ on a défini un opérateur unitaire U_τ (à une constante multiplicative près). Ces opérateurs vérifient $U_\tau U_\sigma = U_{\tau\sigma}\rho(\tau, \sigma)$ où ρ est un complexe de module 1. Il n'est pas possible de choisir les opérateurs U_τ de sorte que les nombres ρ soient tous égaux à 1, (on aurait alors une représentation de $SL(2, \mathbb{R})$), mais on peut en revanche construire une représentation du revêtement double de $SL(2, \mathbb{R})$ (le groupe métaplectique) appelée représentation de Segal-Shale-Weil (cf G.Lion, M.Vergne [48]).

La représentation de l'algèbre de Lie associée est composée d'opérateurs de la forme $\mathcal{P}(P, Q)$, où $\mathcal{P}$ est un polynôme homogène de degré 2 (voir aussi le chapitre I, paragraphes 15 et suivants, de V.Guillemin, S.Sternberg [37]).

Chapitre 4 − L'espace de Fock dont il est question ici est l'espace de Fock symétrique, ou bosonique dans la terminologie physique. Il existe aussi une notion d'espace de Fock antisymétrique (appelé aussi espace de Fock fermionique cf Meyer [50]), et d'espace de Fock complet (voir Speicher [82]), sur lesquels sont définis des opérateurs de création et d'annihilation qui satisfont de nouvelles relations de commutation. On peut également définir des déformations des relations de commutation (25) comme l'ont fait Bozejko et Speicher [15], qui permettent "d'interpoler" entre les relations des bosons et celles des fermions.

L'espace de Fock complet est très lié à la théorie de Voiculescu (cf [85]) des variables aléatoires libres qui a eu des applications remarquables dans la théorie des algèbres d'opérateurs.

La décompositions en chaos de l'espace L^2 du mouvement brownien est dûe à N.Wiener [88]. Ce résultat a été étendu aux processus de Lévy par K.Itô [44]. Pour les martingales normales, la théorie des décompositions chaotiques en est encore à ses débuts, les principaux résultats dans ce domaine sont dûs à M.Emery [26], [27], [28], (voir aussi le chapitre de [23] consacré à ces questions).

Chaque interprétation probabiliste de l'espace de Fock permet de considérer ses éléments commes des variables aléatoires, que l'on peut en particulier multiplier entre elles. Il existe des formules explicites pour le résultat d'une telle multiplication dans les cas des interprétations brownienne et poissonienne que l'on peut trouver dans [50]. Ces formules s'écrivent agréablement lorsqu'on utilise la notation de Guichardet pour l'espace de Fock ([38], [50]).

Chapitre 5 — Nous avons considéré l'espace de Fock construit sur $L^2(\mathbb{R}_+)$, qui possède une structure naturelle de produit tensoriel continu. On aurait pu, de façon plus intrinsèque se donner un espace de Hilbert H muni d'une famille croissante $(H_t)_{t \in \mathbb{R}_+}$ de sous-espaces. Ce point de vue est adopté dans le livre de Parthasarathy [60].

Une notion importante, que nous n'avons pas introduite ici est celle de martingale sur l'espace de Fock. Cette notion est discutée dans [50] et [60], citons aussi les références importantes [63] et [64] où des théorèmes de représentation des martingales analogues au théorème de représentation prévisible d'Itô sont démontrés. La situation est toutefois loin d'être aussi simple que dans le cas classique comme le montre un contre exemple de J.L.Journé [46].

La construction de l'intégrale stochastique non-commutative donnée dans le texte est inspirée d'un article de J.M.Lindsay [47]. Elle nécessite une certaine dose d'analyse fonctionnelle mais a l'avantage d'éviter le recours à des approximations de processus par des processus simples. Une autre construction, qui utilise la résolution d'une équation différentielle, a été donnée par Meyer dans [51].

Le lien entre intégrale non-commutative et intégrale de Skorokhod a été exploité par Belavkin [6], qui s'est inspiré des travaux de Maassen sur les opérateurs représentables par des noyaux ([49]).

L'inégalité sur les intégrales stochastiques non-commutatives est dûe à J.L.Journé (cf [50]). Nous l'avons établie ici en partant d'une inégalité semblable sur les intégrales de Skorokhod. Le problème de savoir si un processus de la forme $u_t v_t$, avec u_t mesurable par rapport au passé et v_t par rapport au futur, est intégrable au sens de Skorokhod a été étudié, entre autres, par M.Jolis et M.Sanz [45], qui ont donné des conditions suffisantes pour que ce soit le cas. Leurs calculs ont une forme assez semblable à celle employée dans le cours, mais dans un cadre plus général (la forme de v_t est y plus générale que dans le texte).

Le formalisme de Belavkin présenté à la fin du paragraphe 5.4 est développé dans [8] sous le nom d'algèbre d'Itô. Pour une application à la structure des fonctions de type positif, voir [7].

Les martingales d'Azéma ont été étudiées par Emery [26]. Ce sont les premières martingales qui ne soient pas des processus à accroissement indépendants pour lesquels on a démontré la propriété de représentation chaotique, néanmoins, un peu plus tard, M.Schürmann a montré que l'on pouvait considérer une martingale d'Azéma comme une composante d'un processus non-commutatif à accroissements indépendants (voir [77]).

Parallèlement au calcul stochastique bosonique présenté ici, il existe un calcul stochastique fermionique ([5]) et un calcul stochastique libre ([82]). Ces calculs stochastiques peuvent être unifiés comme l'on montré Parthasarathy et Sinha [65] (voir aussi [41]).

Chapitre 6 — Pour la notion d'espérance conditionnelle en probabilités quantique, on peut lire D.Petz [67].

La construction de la dilatation présentée dans le paragraphe 6.2 se trouve dans [11] et [62]. Pour des exemples explicites de processus ainsi obtenus, voir [12], [13]. Le processus de Spin, considéré comme un processus à valeurs dans le dual de $SU(2)$ a été considéré par Eymard et Roynette [32].

Des réalisations en temps continu de ces processus, en lien avec le calcul stochastique non-commutatif ont été obtenues par Hudson et Parthasarathy [42].

Le problème de l'existence d'une solution unitaire à l'équation (99) lorsque les coefficients L^ε ne sont pas bornés et la multiplicité est infinie fait l'objet de recherches très actives. Récemment, des résultats importants ont été obtenus, notamment par Chebotarev, Fagnola, Mohari (voir l'exposé de Meyer au séminaire Bourbaki [53]). Ce problème est crucial pour la construction des processus de Markov au moyen des équations différentielles stochastiques non-commutatives. Le problème de l'existence et de l'unicité d'une solution unitaire est un analogue non-commutatif de la théorie des frontières pour les chaînes de Markov (cf [54]).

Le théorème 9 est le résultat principal de l'article [40], et son obtention était la motivation initiale du calcul stochastique non-commutatif. Le semi-groupe e^{tA} est complètement positif, et réciproquement, le théorème de Gorini, Kossakowski, Sudarshan, Lindblad (démontré dans [60]) énonce que tout semi-groupe complètemment positif sur $\mathcal{B}(H)$, uniformément continu a la forme e^{tB} où B est un opérateur qui s'exprime comme une somme d'opérateurs du type de $A^{\cdot}$.

En imitant la proposition 23 on peut se poser le problème de l'existence de morphismes $j_t : \mathcal{B}(H) \to \mathcal{W}$ satisfaisant l'équation

$$dj_t(x) = \sum_\varepsilon j_t(A^\varepsilon x) da_t^\varepsilon$$

où les A^ε sont des applications linéaires de $\mathcal{B}(H)$ dans $\mathcal{B}(H)$. Ces morphismes forment alors une *diffusion quantique*, et les A^ε sont des analogues des opérateurs différentiels qui apparaissent dans l'équation (110). La formule d'Itô, associée aux condition que les j_t soient multiplicatifs, préservent l'involution et l'identité, entraînent des conditions nécessaires algébriques sur les A^ε pour l'existence des j_t. Si ces conditions sont remplies, le problème de l'existence d'une solution a été résolu affirmativement si les A^ε sont des opérateurs bornés (cf [30], [31]). Dans le cas d'opérateurs A^ε non-bornés, on n'a que des résultats partiels ([33]).

Le contenu du paragraphe 6.5 est inspiré de [51], VI.3, et [66].

Citons encore d'autres travaux sur des thèmes voisins des diffusions quantiques, par Applebaum [3], [4], Davies et Lindsay [19], Sauvageot [75], [76].

Références

[1] L.Accardi, A.Frigerio, J.T.Lewis, Quantum Stochastic processes, *Publ. R.I.M.S. Kyoto*, 18, 1982, 94-133.

[2] L.Accardi, W.von Waldenfels Eds., Quantum Probability and applications, I-V Lect. Notes in Maths. Springer 1055, 1136, 1303, 1325, 1442, VI- World Scientific, Singapore.

[3] D.Applebaum, Unitary evolutions and horizontal lifts in quantum stochastic calculus, *Comm. Math. Phys.*, 140, 1991, 63-80.

[4] D.Applebaum, An operator theoretic approach to stochastic flows on manifolds, *Séminaire de Probabilités XXVI*, Lect. Notes in Maths. Springer 1526, 1992, 514-532.

[5] C.Barnett, R.F.Streater, I.F.Wilde, The Ito-Clifford integral I, *Jour. Funct. An.*, 48, 1982, 172-212.

[6] V.P.Belavkin, A quantum non-adapted Ito formula and stochastic analysis in Fock scale, *Jour. Funct. An.*, 102, 1991, 414-447.

[7] V.P.Belavkin, Chaotic states and stochastic integration in quantum systems, *Russian Math. Surv.* 14:1 1992, 53-116.

[8] V.P.Belavkin, The unified ito formula has the pseudo-Poisson structure, *preprint n° 98* Centro Vito Volterra, Universita degli studi di Roma II, 1992.

[9] B.V.R.Bhat, K.R.Parthasarathy, Markov dilations of non-conservative dynamical semi-groups and a quantum boundary theory, *preprint*, 1992.

[10] P.Biane, Chaotic representation for finite Markov chains, *Stochastics*, 30, 1990, 61-68.

[11] P.Biane, Marches de Bernoulli quantiques, *Séminaire de Probabilités XXIV*, Lect. Notes in Maths. Springer 1426, 1990, 329-344.

[12] P.Biane, Quantum random walks on the dual of SU(n), *Prob. Th. and Rel. Fields*, 89, 1991, 117-129.

[13] P.Biane, Minuscule weights and random walks on lattices, *Quantum probability and applications*, VII, World Scientific, Singapore, 1992, 51-65.

[14] N.Bourbaki, Groupes et algèbres de Lie, Chapitre 9, Hermann, Paris, 1969.

[15] M.Bozejko, R.Speicher, An example of generalized brownian motion, *Comm. Math. Phys.*, 137, 1991, 519-531.

[16] T.Chihara, An introduction to orthgonal polynomials, Gordon and Breach, New York, 1978.

[17] K.L.Chung, R.J.Williams, Introduction to stochastic integration, *Progress in Mathematics*, Birkhaüser, 1983.

[18] E.B.Davies, Quantum theory of open systems, Academic Press, 1976.

[19] E.B.Davies, J.M.Lindsay, Non-commutative Markov semi-groups, *preprint*, 1990.

[20] C.Dellacherie, P.A.Meyer, Probabilités et potentiel, Chapitre I à IV, Hermann, Paris, 1975.

[21] C.Dellacherie, P.A.Meyer, Probabilités et potentiel, Chapitre V à VIII, Hermann, Paris, 1980.

[22] C.Dellacherie, P.A.Meyer, Probabilités et potentiel, Chapitre IX à XI, Hermann, Paris, 1983.

[22] C.Dellacherie, P.A.Meyer, B.Maisonneuve, Probabilités et potentiel, Chapitre XVII à XXIV, Hermann, Paris, 1993.

[24] J.Dixmier, Les algèbres d'opérateurs dans l'espace hilbertien, (Algèbres de von Neumann), Gauthier-Villars, Paris, 1957.

[25] J.Dixmier, Les C*-algèbres et leurs représentations, Gauthier-Villars, Paris, 1964.

[26] M.Emery, On the Azéma martingales, *Séminaire de Probabilités XXIII*, Lect. Notes in Maths. Springer 1372, 1990, 66-87.

[27] M.Emery, Quelques cas de représentation chaotique, *Séminaire de Probabilités XXIV*, Lect. Notes in Maths. Springer 1426, 1991, 10-23.

[28] M.Emery, On the chaotic representation property for martingales, *preprint* 1993.

[29] D.E.Evans, J.T.Lewis, Dilations of dynamical semi-groups, *Comm. Math. Phys.*, 50, 1976, 219-228.

[30] M.Evans, Existence of quantum diffusions, *Prob. Th. and Rel. Fields*, 81, 1989, 473-483.

[31] M.Evans, R.L.Hudson, Multidimensionnal quantum diffusions, *Quantum probability and applications*, III, Lect. Notes in Maths. Springer 1303, 1988, 69-88.

[32] P.Eymard, B.Roynette, Marches aléatoires sur le dual de SU(2), *Marches aléatoires sur les groupes*, Lect. Notes in Maths. Springer 624, 1977.

[33] F.Fagnola, K.B.Sinha, Quantum flows with unbounded structure maps and finite degrees of freedom, *à paraître dans Jour. Lond. Math. Soc.*

[34] W.Feller, An introduction to probability theory and its applications, Vol I, John Wiley & Sons, New York 1977.

[35] R.P.Feynman, R.B.Leighton, M.Sands, The Feynman Lectures on physics, Vol III, Addison Wesley, Reading Mass. 1965.

[36] B.Gaveau, P.Trauber, L'intégrale stochastique comme opérateur de divergence dans l'espace fonctionnel, *Jour. Funct. An.* 46, 1982, 230-238.

[37] V.Guillemin, S.Sternberg, Symplectic techniques in Physics, Cambridge University Press, 1984.

[38] A.Guichardet, Symmetric Hilbert space and related topics, *Lect. Notes in Maths. Springer* 261, 1970.

[39] R.Halmos, Introduction to Hilbert space, Chelsea, New York, 1951.

[40] R.L.Hudson, K.R.Parthasarathy, Quantum Ito's formula and stochastic evolutions, *Comm. Math. Phys.*, 93, 1984, 301-323.

[41] R.L.Hudson, K.R.Parthasarathy, Unification of Fermion and Boson stochastic calculus, *Comm. Math. Phys.*, 115, 1988, 47-53.

[42] R.L.Hudson, K.R.Parthasarathy, Casimir chaos in Boson Fock space, *preprint*, 1993.

[43] N.Ikeda, S.Watanabe, Stochastic differential equations and diffusion processes, North-Holland, Kodansha, 1981.

[44] K.Itô, Multiple Wiener integrals, *Jour. Math. Soc. Japan.* 3, 1951, 157-169.

[45] M.Jolis, M.Sanz, Integrator properties of the Skorokhod integral, *Stochastics,*, 41, 1992, 163-176.

[46] J.L.Journé, P.A.Meyer, Une martingale d'opérateurs bornés non représentable en intégrale stochastique, *Séminaire de Probabilités XX*, Lect. Notes in Maths. Springer 1204, 1986, 313-316.

[47] M.Lindsay, Quantum and non-causal stochastic calculus, *à paraître dans Prob. Theory and rel. Fields.*

[48] G.Lion, M.Vergne, The Weil representation, Maslov index, and theta series, Progress in Mathematics, Vol 6, Birkhaüser, 1980.

[49] H.Maassen, Quantum Markov processes on Fock space described by integral kernels, *Quantum probability and applications*, II, Lect. Notes in Maths. Springer 1136, 1985, 361-374.

[50] P.A.Meyer, Eléments de Probabilités quantiques, *Séminaire de Probabilités XX*, Lect. Notes in Maths. Springer 1204, 1986, 186-312.

[51] P.A.Meyer, Quantum Probability for Probabilists, Lect. Notes in Maths. Springer 1538, 1993.

[52] P.A.Meyer, Approximation de l'oscillateur harmonique (d'après L.Accardi et A.Bach), *Séminaire de Probabilités XXIII*, Lect. Notes in Maths. Springer 1372, 1990, 175-182.

[53] P.A.Meyer, Progrès récent en calcul stochastique quantique, *Séminaire Bourbaki*, exposé 761, 1992.

[54] A.Mohari, K.R.Parthasarathy, A quantum probabilistic analogue of Feller's condition for the existence of unitary markovian cocycles in Fock space, *I.S.I. preprint*, 1992.

[55] E.Nelson, Analytic vectors, *Ann. Math.* 70, 1959, 572-615.

[56] J.von Neumann, Mathematical Foundations of Quantum Mechanics, Princeton University Press, 1951.

[57] J.Neveu, Processus Aléatoires Gaussiens, Presses de l'Université de Montréal, 1968.

[58] J.Neveu, Processus Ponctuels, *Ecole d'élé de Probabilités de Saint-Flour VI*, Lect. Notes in Maths. Springer 598, 1976, 250-445.

[59] D.Nualart, Non-causal stochastic integrals and calculus, *Stochastic analysis and related topics,* Lect. Notes in Maths. Springer 1316, 1986, 80-129.

[60] K.R.Parthasarathy, An introduction to quantum stochastic calculus, Monographs in mathematics, Vol 85, Birkhäuser, 1992.

[61] K.R.Parthasarathy, Azéma martingales and quantum stochastic calculus, Proc. R.C.Bose Symposium, Wiley Eastern, 1990, 551-569.

[62] K.R.Parthasarathy, A generalized Biane's process, *Séminaire de Probabilités XXIV*, Lect. Notes in Maths. Springer 1426, 1990, 345-348.

[63] K.R.Parthasarathy, K.B.Sinha, Representation of bounded martingales in Fock space, *Jour. Funct. An.*, 67, 1986, 126-151

[64] K.R.Parthasarathy, K.B.Sinha, Representation of a class of bounded martingales, II, *Quantum probability and applications*, III, Lect. Notes in Maths. Springer 1303, 1988, 232-250.

[65] K.R.Parthasarathy, K.B.Sinha, Unification of quantum noise processes in Fock spaces, *Quantum probability and applications*, VI, World Scientific, Singapore, 1991.

[66] K.R.Parthasarathy, K.B.Sinha, Markov chains as Evans-Hudson diffusions in Fock space, *Séminaire de Probabilités XXIV*, Lect. Notes in Maths. Springer 1426, 1990, 362-369.

[67] D.Petz, Conditionnal expectations in quantum probability, *Quantum probability and applications*, III, Lect. Notes in Maths. Springer 1303, 1988, 251-260.

[68] J.Pitman, One dimensionnal brownian motion and the three dimensionnal Bessel process, *Adv. Appl. Prob.*, 7, 1975, 511-526.

[69] P.Protter, Stochastic integration and differential equations, a new approach, Springer, 1990.

[70] M.Reed, B.Simon, Methods of modern mathematical physics, II, Fourier analysis and self-adjointness, Academic press, 1970.

[71] S.Reynaud, Introduction à la réduction du bruit quantique, *Ann. Phys. Fr.* 15, 1990, 63-162.

[72] W.Rudin, Functionnal analysis, Mac Graw Hill, 1973.

[73] J.de Sam Lazaro, P.A.Meyer, Méthodes de martingales et théorie des flots, *Zeit. f. Wahr.* 18, 1971, 116-140.

[74] J.L.Sauvageot, Markov quantum semi-groups admit covariant Markov C^*-dilations, *Comm. Math. Phys.* 106, 1986, 91-103.

[75] J.L.Sauvageot, Quantum Dirichlet forms, differential calculus and semi-groups, *Quantum probability and applications*, V, Lect. Notes in Maths. Springer 1442, 1990, 334-346.

[76] J.L.Sauvageot, Semi-groupe de la chaleur transverse sur la C^*-algèbre d'un feuilletage riemannien, C.R.A.S. 310, Série I, 1990, 531-536.

[77] M.Schürmann, The Azéma martingales as components of quantum independent increment processes, *Séminaire de Probabilités XXV*, Lect. Notes in Maths. Springer 1485, 1991, 24-30.

[78] J.P.Serre, Algèbres de Lie semi-simples complexes, Benjamin, New York, 1966.

[79] D.Shale, Linear symmetries of free boson fields, *Trans. Am. Math. Soc.* 103, 1962, 149-167.

[80] A.V.Skorokhod, On a generalisation of a stochastic integral, *Teor. Ver.* 20, 1975, 219-233.

[81] R.Speicher, A new example of "independance" and "white noise", *Prob. Th. and Rel. Fields*, 84, 1990, 141-159.

[82] R.Speicher, Stochastic integration on the full Fock space with the help of a fernel calculus, *Publ. R.I.M.S. Kyoto*, 27, 1991, 149-184.

[83] N.J.Vilenkin, Special functions and the theory of group representations, *Translations of the A.M.S.*, 22, 1968.

[84] G.F.Vincent-Smith, Dilation of a dissipative quantum dynamical system into a quantum Markov process, *Proc. Lond. Math. Soc.* 49, 1984, 58-72.

[85] D.Voiculescu, Free non-commutative random variables, random matrices and the II_1 factors of free groups, *Quantum probability and applications*, VI, World Scientific, Singapore, 1991, 473-487.

[86] W.von Waldenfels, The Markov process of total spin, *Séminaire de Probabilités XXIV*, Lect. Notes in Maths. Springer 1426, 1990, 357-361.

[87] S.Watanabe, Stochastic differential equations and Malliavin calculus, Lectures on Mathematics and Physics, 73, Tat Institute of Fundamental Research, 1984.

[88] N.Wiener, The homogeneous chaos, *Amer. Jour. Math.* 55, 1938, 897-936.

CNRS, Laboratoire de Probabilités
Université Paris 6, Tour 56 3e étage
4 place Jussieu 75252 PARIS Cedex 05

TEN LECTURES

ON PARTICLE SYSTEMS

Rick DURRETT

Preface. These lectures were written for the 1993 St. Flour Probability Summer School. Their aim is to introduce the reader to the mathematical techniques involved in proving results about interacting particle systems. Readers who are interested instead in using these models for biological applications should instead consult Durrett and Levin (1993).

In order that our survey is both broad and has some coherence, we have chosen to concentrate on the problem of proving the existence of nontrivial stationary distributions for interacting particle systems. This choice is dictated at least in part by the fact that we want to make propaganda for a general method of solving this problem invented in joint work with Maury Bramson (1988): comparison with oriented percolation. Personal motives aside, however, the question of the existence of nontrivial stationary distributions is the first that must be answered in the discussion of any model.

Our survey begins with an overview that describes most of the models we will consider and states the main results we will prove, so that the reader can get a sense of the forest before we start investigating the individual trees in detail. In Section 2 we lay the foundations for the work that follows by proving an existence theorem for particle systems with translation invariant finite range interactions and introducing some of the basic properties of the resulting processes. In Section 3 we give a second construction that applies to a special class of "additive" models, that makes connections with percolation processes and that allows us to define dual processes for these models.

The general method mentioned above makes its appearance in Section 4 (with its proofs hidden away in the appendix) and allows us to prove a very general result about the existence of stationary distributions for attractive systems with state space $\{0,1\}^S$. The comparison results in Section 4 are the key to our treatment of the threshold contact and voter models in Section 5, the cyclic systems in Section 6, the long range contact process in Section 7, and the predator prey system in 9.

In Section 7 we explore the first of two methods for simplifying interacting particle systems: assuming that the range of interaction is large. In Section 8 we meet the second: superimposing particle motion at a fast rate. The second simplification leads to a connection with reaction diffusion equations which we exploit in Section 9 to prove the existence of phase transitions for predator prey systems.

The quick sketch of the contents of these lectures in the last three paragraphs will be developed more fully in the overview. Turning to other formalities, I would like to thank the organizers of the summer school for this opportunity to speak and write about my favorite subject. Many of the results presented here were developed with the support of the National Science Foundation and the Army Research Office through the Mathematical Science Institute at Cornell University. During the Spring semester of 1993, I gave 10 one and a half hour lectures to practice for the summer school and to force myself to get the writing done on time. You should be grateful to the eight people who attended this dress rehearsal: Hassan Allouba, Scott Arouh, Itai Benjamini, Carol Bezuidenhout, Elena Bobrovnikova, Sungchul Lee, Gang Ma, and Yuan-Chung Sheu, since their suffering has lessened yours.

Although it is not yet the end of the movie, I would like to thank the supporting cast now: Tom Liggett, who introduced me to this subject; Maury Bramson, the co-discoverer of the comparison method and long range limits, to whom I turn when my problems get too hard; David Griffeath, my electronic colleague who introduced me (and the rest of the world) to the beautiful world of the Greenberg Hastings and cyclic cellular automata; Claudia Neuhauser, my former student who constantly teachs me how to write; and Ted Cox, with whom I have written some of my best papers. The field of interacting particle systems has grown considerably since Liggett's 488 page book was published in 1985, so it is inevitable that more is left out than is covered in these notes. The most overlooked researcher in this treatment is Roberto Schonmann whose many results on the contact process, bootstrap percolation, and metastability in the Ising model did not fit into our plot.

1. Overview

   In an interacting particle system, there is a countable set of spatial locations S called *sites*. In almost all of our applications $S = \mathbf{Z}^d$, the set of points in d dimensional space with integer coordinates. Each site can be in one of a finite set of *states* F, so the state of the system at time t is $\xi_t : S \to F$ with $\xi_t(x)$ giving the state of x. To describe the evolution of these models, we specify an *interaction neighborhood*

$$\mathcal{N} = \{z_0, z_1, \ldots z_k\} \subset \mathbf{Z}^d$$

with $z_0 = 0$ and define *flip rates*

$$c_i(x, \xi) = g_i(\xi(x + z_0), \xi(x + z_1), \ldots, \xi(x + z_k))$$

In words, the state of x flips to i at rate $c_i(x, \xi)$ when the state of the process is ξ. In symbols, if $\xi_t(x) \neq i$ then

$$\frac{P(\xi_{t+s}(x) = i | \xi_t = \xi)}{s} \to c_i(x, \xi) \quad \text{as } s \to 0$$

The formula for c_i indicates that our interaction is *finite range*, i.e., the flip rates depend only on the state of x and of a finite number of neighbors; and *translation invariant*, i.e., the rules applied at x are just a translation of those applied at 0.

   To explain what we have in mind when making these definitions, we now describe two famous concrete examples. In this section and throughout these lectures (with the exception of Sections 2 and 3) we will suppose that

$$\mathcal{N} = \{x : \|x\|_p \leq r\}$$

Here $r \geq 1$ is the *range* of the interaction and $\|x\|_p$ is the usual L^p norm on $\mathbf{R}^d$. That is, $\|x\|_p = (x_1^p + \ldots + x_d^p)^{1/p}$ when $1 \leq p < \infty$ and $\|x\|_\infty = \sup_i |x_i|$. In most of our models the flip rates are based on the number of neighbors in state i, so we introduce the notation:

$$n_i(x, \xi) = |\{z \in \mathcal{N} : \xi(x + z) = i\}|$$

where $|A|$ is the number of points in A.

Example 1.1. The basic contact process. To model the spread of a plant species we think of each site x as representing a square area in space with $\xi_t(x) = 0$ if that area is vacant and $\xi_t(x) = 1$ if there is a plant there, and we formulate the dynamics as follows:

$$c_0(x, \xi) = \delta \quad \text{if } \xi(x) = 1$$
$$c_1(x, \xi) = \lambda n_1(x, \xi) \quad \text{if } \xi(x) = 0$$

In words, plants die at rate δ independent of the state of their neighbors, while births at vacant sites occur at a rate proportional to the number of occupied neighbors. Note that flipping to i has no effect when $\xi(x) = i$ so the value of $c_i(x, \xi)$ on $\{\xi(x) = i\}$ is irrelevant and we could delete the qualifying phrases "if $\xi(x) = 1$" and "if $\xi(x) = 0$" if we wanted to.

Example 1.2. The basic voter model. This time we think of the sites in Z^d as representing an array of houses each of which is occupied by one individual who can be in favor of ($\xi_t(x) = 1$) or against ($\xi_t(x) = 0$) a particular issue or candidate. Our simple minded voters change their opinion to i at a rate that is equal to the number of neighbors with that opinion. That is,

$$c_i(x, \xi) = n_i(x, \xi)$$

The first question to be addressed for these models is:

Do the rates specify a unique Markov process?

There is something to be proved since there are infinitely many sites and hence no first jump, but for our finite range translation invariant models, a result of Harris (1972) allows us to easily show that the answer is Yes. (See Section 2.) The main question we will be interested in is:

When do interacting particle systems have a nontrivial stationary distributions?

To make this question precise we need a few definitions. The state space of our Markov process is F^S, the set of all functions $\xi : S \to F$. We let $\mathcal{F} = $ all subsets of F and equip F^S with the usual product σ-field $\mathcal{F}^S$, which is generated by the *finite dimensional sets*

$$\{\xi(y_1) = i_1, \ldots, \xi(y_k) = i_k\}$$

So any measure π on $\mathcal{F}^S$ can be described by giving its *finite dimensional distributions*

$$\pi(\xi(y_1) = i_1, \ldots, \xi(y_k) = i_k)$$

As in the theory of Markov chains, π is said to be a *stationary distribution* for the process if when we start from an initial state ξ_0 with distribution π (i.e., $\pi(A) = P(\xi_0 \in A)$ for $A \in \mathcal{F}^S$) then ξ_t has distribution π for all $t > 0$. Since our dynamics are translation invariant, we will have a special interest in stationary distributions that are *translation invariant*, i.e., ones in which the probabilities $\pi(\xi(x + y_1) = i_1, \ldots, \xi(x + y_k) = i_k)$ do not depend upon x.

To explain the term "nontrivial" we note that in Example 1.1 the "all 0" state ($\xi(x) \equiv 0$) and in Example 1.2 for any i the all i state are *absorbing states*. That is, once the process enters these states it cannot leave them. If S were finite this fact (and enough irreducibility, which is present in Examples 1.1 and 1.2) would imply that all stationary distributions were *trivial*, i.e., concentrated on absorbing states. However, when S is infinite this argument fails and indeed, as the next few results show it is possible to have a nontrivial stationary distributions.

Theorem 1. Consider the basic contact process with $\mathcal{N} = \{x : \|x\|_p \leq r\}$ with $r \geq 1$. If $\lambda|\mathcal{N}| \leq \delta$ then there is only the trivial stationary distribution. If $\delta/\lambda < \delta_0$ then there is a nontrivial translation invariant stationary distribution.

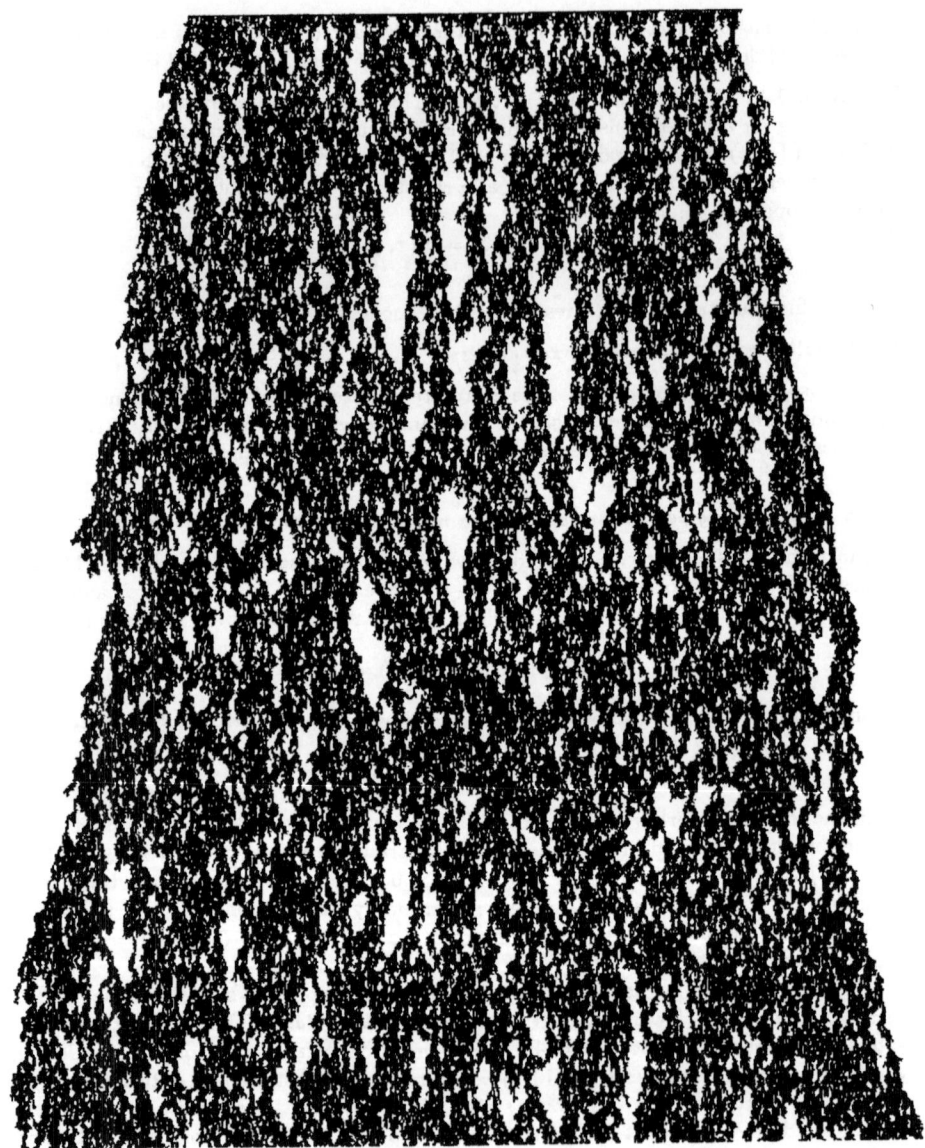

Figure 1.1. Nearest neighbor contact process in $d = 1$ with $\lambda = 2$.

The first result is easy to see. If the contact process has k particles then the number drops to $k - 1$ at rate δk and increases to $k + 1$ at rate $\leq \lambda |\mathcal{N}|$ with the upper bound achieved when all particles are isolated (i.e., no two particles are neighbors). The reader should attempt to prove the converse before we hit it with our sledgehammer in Section 4. By a simple comparison that you will learn about in Section 2, it is enough to prove the result when $\mathcal{N} = \{x : \|x\|_1 = 1\}$ and $d = 1$. A simulation of this case with $\lambda = 2$ is given in Figure 1.1. A result of Holley and Liggett (1978) implies that in this situation there is a nontrivial stationary distribution. In our simulation we have started with the interval $[180, 540]$ occupied at time 0 at the top of the page. As time runs down the page from 0 to 720, it is clear that the region occupied by particles is growing linearly, as predicted by a result of Durrett (1980).

Turning to the voter model, the classic paper of Holley and Liggett (1975) tells us that

Theorem 2A. *Clustering* occurs in $d \leq 2$. That is, for any ξ_0 and $x, y \in \mathbf{Z}^d$ we have

$$P(\xi_t(x) \neq \xi_t(y)) \to 0 \text{ as } t \to \infty$$

Theorem 2B. Let ξ_t^θ denote the process starting from an initial state in which the events $\{\xi_0^\theta(x) = 1\}$ are independent and have probability θ. In $d \geq 3$ as $t \to \infty$, $\xi_t^\theta \Rightarrow \xi_\infty^\theta$, a translation invariant stationary distribution in which $P(\xi_\infty^\theta(x) = 1) = \theta$.

Here $\Rightarrow$ denotes *weak convergence*, which in this setting is just convergence of finite dimensional distributions. That is, for any $x_1, \dots x_m \in \mathbf{Z}^d$ and $i_1, \dots, i_m \in \{1, 2 \dots, \kappa\}$ we have

$$P(\xi_t^\theta(x_1) = i_1, \dots \xi_t^\theta(x_m) = i_m) \to P(\xi_\infty^\theta(x_1) = i_1, \dots \xi_\infty^\theta(x_m) = i_m)$$

We will say that *coexistence* occurs if there is a translation invariant stationary distribution in which each of the possible states in F has positive density. Theorems 2A and 2B say that in the voter model coexistence is possible in $d \geq 3$ but not in $d \leq 2$. We will see in Section 3 that this is a consequence of the fact that if we take two independent random walks with jumps uniformly distributed on $\mathcal{N}$ then they will hit with probability 1 in $d \leq 2$ but with probability < 1 in $d \geq 3$.

Figure 1.2 gives a simulation of a voter model with five opinions on $\{0, 1, \dots, 119\}^2$. Here and in the next six simulations in this section, we use periodic boundary conditions. That is, sites on the top row are neighbors of those on the bottom row, and those on the left edge are neighbors of those on the right edge. We started at time 0 by assigning a randomly chosen opinion to each site. Figure 1.2 shows the state at time 500 suggesting that the clustering asserted in Theorem 2A occurs very slowly. Results of Cox (1988) imply that the expected time for our system to reach consensus is about

$$4 \ln(5/4) \cdot \frac{2}{\pi} (120)^2 \ln 120 = 39,173$$

The conclusions just derived for the voter depend on the fact that the flip rates are linear. Nonlinear flip rates can produce quite different behavior:

Figure 1.2. Five opinion two dimensional voter model at time 500

Example 1.3. The threshold voter model. Cox and Durrett (1991) introduced a modification of the voter model in which

$$c_i(x, \xi) = \begin{cases} 1 & \text{if } n_i(x, \xi) \geq \theta \\ 0 & \text{if } n_i(x, \xi) < \theta \end{cases}$$

In words, these voters change their opinion at rate 1 if at least θ neighbors disagree with them. This change in the rules changes the behavior of the model drastically.
   We start with the case $\theta = 1$:

Theorem 3A. If $d = 1$ and $\mathcal{N} = \{-1, 1\}$ then clustering occurs.

Theorem 3B. In all other cases (recall we supposed that $\mathcal{N} = \{z : \|z\|_p \leq r\}$ with $r \geq 1$) we have coexistence. That is, there is a nontrivial translation invariant stationary distribution μ_{12} in which 1's and 2's each have density 1/2.

Here as in many other cases, the one dimensional nearest neighbor case is an exception. Cox and Durrett (1991) proved Theorem 3A and that coexistence occurs in some cases (e.g., $d = 1$ and $r \geq 7$) but the sharp Theorem 3B is due to Liggett (1992). Note that in the threshold voter model coexistence occurs in all but one case, while in the basic voter model coexistence occurs only in $d \geq 3$. A second difference is that when coexistence occurs the basic voter model has a one parameter family of nontrivial stationary distributions constructed in Theorem 2B but we believe

Conjecture 3C. When coexistence occurs in the threshold one voter model there is a unique spatially ergodic translation invariant stationary distribution in which 1's and 2's have positive density.

Here, we say that π on F^S is *spatially ergodic* if under π the family of random variables $\{\xi(x) : x \in \mathbf{Z}^d\}$ is an ergodic stationary sequence, i.e., the σ-field of events invariant under all spatial shifts is trivial. We need the assumption of spatial ergodicity to rule out nontrivial convex combinations

$$a\mu_1 + b\mu_2 + (1 - a - b)\mu_{12}$$

where μ_i is the point mass on the all i state, and μ_{12} is the measure constructed in Theorem 3B. In general, the set of translation invariant stationary distributions for an interacting particle system is a convex set and in most examples, the extreme points of the set are the stationary distributions that are spatially ergodic. However, there is no general result that shows this is true. See Problem 7 on page 178 of Liggett (1985).
   While the threshold 1 case is fairly well understood, there are many open problems concerning higher thresholds. To illustrate these we observe that computer simulations suggest

Conjecture 3D. For the Moore neighborhood $\mathcal{N} = \{z : \|z\|_\infty = 1\}$ in $d = 2$ the threshold voter model has the following behaviors

Figure 1.3. Threshold 2 voter model, Moore neighborhood

Figure 1.4. Threshold 3 voter model, Moore neighborhood

coexistence	$\theta = 1, 2$
clustering	$\theta = 3, 4$
fixation	$\theta \geq 5$

Here, *fixation* means that each sites flips only a finite number of times. To see that the last line is a reasonable guess note that an octagon of 1's cannot flip to 0 since each 1 has at most 4 neighbors that are 0

0	0	0	0	0	0
0	0	1	1	0	0
0	1	1	1	1	0
0	1	1	1	1	0
0	0	1	1	0	0
0	0	0	0	0	0

We will prove the result about fixation for $\theta \geq 5$ and coexistence for $\theta = 1$ in Section 5. The other conclusions are open problems. In support of our conjectures we introduce Figures 1.3 and 1.4 which give simulations at time 50 of the case $\theta = 2$ on $\{0, 1, \ldots, 89\}^2$ and $\theta = 3$ on $\{0, 1, \ldots, 179\}^2$ starting from product measure with density 1/2.

Our next two systems model the competition of biological species. We begin with

Example 1.4. The multitype contact process. The set of states is $F = \{0, 1, \ldots, \kappa\}$, where 0 indicates a vacant site and $i > 0$ indicates a site occupied by one plant of type i. The flip rates are linear

$$c_0(x, \xi) = \delta_{\xi(x)}$$
$$c_i(x, \xi) = \lambda_i n_i(x, \xi) \quad \text{if } \xi(x) = 0$$

Here and in what follows the rates we do not mention are 0. Suppose for simplicity that $\kappa = 2$. Neuhauser (1992) has shown

Theorem 4A. Suppose $\delta_1 = \delta_2$ and $\lambda_1 > \lambda_2$. If ξ_0 is translation invariant and has a positive density of 1's then $P(\xi_t(x) = 2) \to 0$.

In words, the species with the higher birth rate wins out ("survival of the fittest"). The following stronger result should be true but Neuhauser's proof relies heavily on the assumption that $\delta_1 = \delta_2$.

Conjecture 4B. Suppose $\lambda_1/\delta_1 > \lambda_2/\delta_2$. If ξ_0 contains infinitely many 1's then $P(\xi_t(x) = 2) \to 0$.

When $\lambda_1 = \lambda_2$ and $\delta_1 = \delta_2$, Neuhauser showed that the multitype contact process behaves like the voter model.

Theorem 4C. *Clustering* occurs for translation invariant initial states in $d \leq 2$. That is, if ξ_0 is translation invariant, then for any $x, y \in \mathbf{Z}^d$, and $1 \leq i < j \leq \kappa$ we have

$$P(\xi_t(x) = i, \xi_t(y) = j) \to 0 \text{ as } t \to \infty$$

Theorem 4D. Let ξ_t^θ denote the process starting from an initial state in which the events $\{\xi_0^\theta(x) = i\}$ are independent and have probability θ_i with $\theta_2 = 1 - \theta_1$. In $d \geq 3$, as $t \to \infty$, $\xi_t^\theta \Rightarrow \xi_\infty^\theta$, a translation invariant stationary distribution in which

$$P(\xi_t(x) = i) = \begin{cases} 1 - \rho & \text{when } i = 0 \\ \rho\theta_i & \text{when } i > 0 \end{cases}$$

where ρ is the equilibrium density of occupied sites in the one type contact process.

The last result is a little disturbing for biological applications. It says that if species compete on an equal footing then coexistence is not possible in $d = 2$ even if the birth and death rates are exactly the same. (This situation may sound unlikely to occur in nature but it occurs, for example, if we look at the competition of *genets* genetically identical individuals of the same species.) Somewhat surprisingly, if species 2 dominates species 1, we get coexistence for an open set of parameter values.

Example 1.5. Successional dynamics. We suppose that the set of states at each site are $0 = $ grass, $1 = $ a bush, $2 = $ a tree and we formulate the dynamics as

$$c_0(x, \xi) = \delta_{\xi(x,\xi)}$$
$$c_1(x, \xi) = \lambda_1 n_1(x, \xi) \quad \text{if } \xi(x) = 0$$
$$c_2(x, \xi) = \lambda_2 n_2(x, \xi) \quad \text{if } \xi(x) \leq 1$$

The title of this example and its formulation are based on the observation that if an area of land is cleared by a fire, then regowth will occur in three stages: first grass appears then small bushes and finally trees, with each species growing up through and replacing the previous one. With this in mind, we allow each type to give birth onto sites occupied by lower numbered types. As in the threshold voter model, the one dimensional nearest neighbor case is an exception.

Theorem 5A. Coexistence is not possible in the one dimensional nearest neighbor case, i.e., $d = 1$, $\mathcal{N} = \{-1, 1\}$.

Conjecture 5B. In all other cases (recall we supposed that $\mathcal{N} = \{z : \|z\|_p \leq r\}$ with $r \geq 1$) we have coexistence for an open set of values $(\delta_1, \lambda_1, \delta_2, \lambda_2)$.

Figure 1.5 shows a simulation of the nearest neighbor model on $\{0, 1, \ldots, 89\}^2$ with parameters $\lambda_1 = 5/4$, $\delta_1 = 1$, $\lambda_2 = 1.9/4$, and $\delta_2 = 1$ run until time 100, which presumably represents the equilibrium state. Sites in state 1 are gray; those in state 2 are black.

Proving that coexistence occurs in the two dimensional nearest neighbor case of this model seems to be a difficult problem, since computer simulations indicate that the open set referred to in Conjecture 5B is rather small. However, if we assume that the range of interaction is large, we can get very accurate results about the coexistence region. Let $\beta_i = \lambda_i |\mathcal{N}|$.

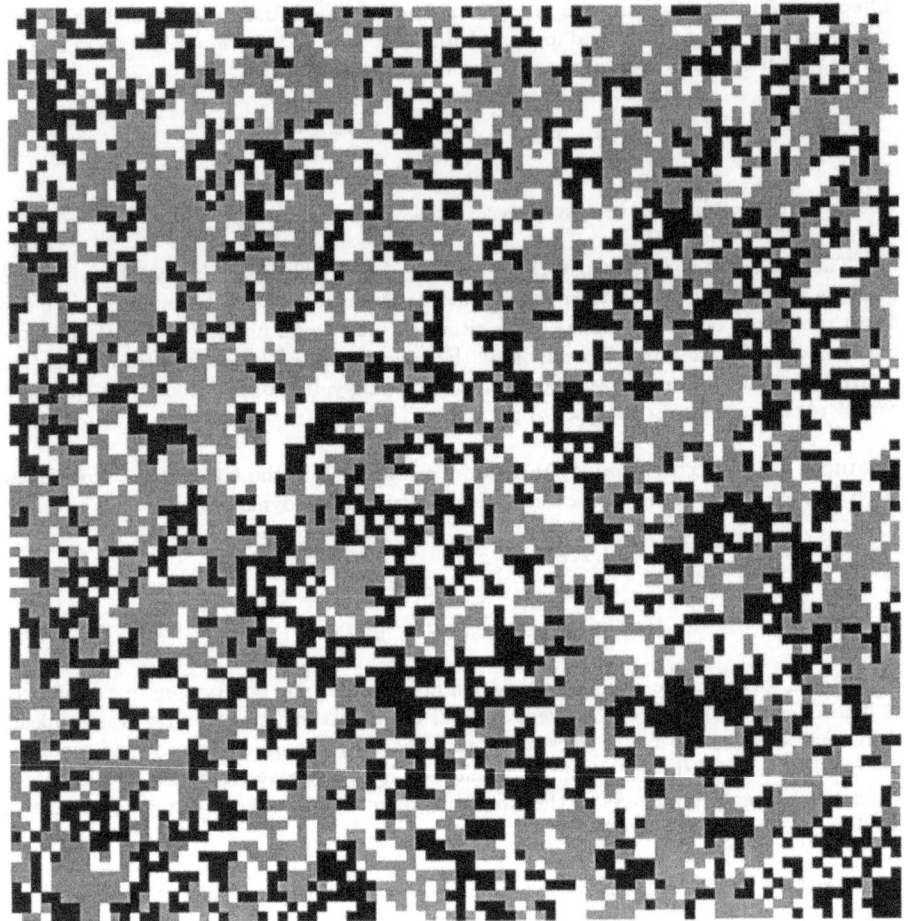

Figure 1.5. Two dimensional nearest neighbor succesional dyanmics, $\beta_1 = 5$, $\beta_2 = 1.9$

Theorem 5C. Suppose that

$$(\star) \qquad \beta_1 \cdot \frac{\delta_2}{\beta_2} > \delta_1 + \beta_2 \cdot \frac{\beta_2 - \delta_2}{\beta_2}$$

If r is large then coexistence occurs.

Theorem 5D. Suppose that

$$\beta_1 \cdot \frac{\delta_2}{\beta_2} < \delta_1 + \beta_2 \cdot \frac{\beta_2 - \delta_2}{\beta_2}$$

If r is large then coexistence is impossible.

Theorems 5A and 5C are due to Durrett and Swindle (1991), while the converse in 5D is due to Durrett and Schinazi (1993). To explain the condition in Theorems 5C and 5D, we begin by observing that if we assumed that $u(t) = P(\xi_t(x) = 2)$ does not depend on x and the states of neighboring sites were independent, then writing $y \sim x$ to denote "y is a neighbor of x"

$$(1.1) \qquad \begin{aligned} \frac{du}{dt} &= -\delta_2 P(\xi_t(x) = 2) + \sum_{y \sim x} \lambda_2 P(\xi_t(x) < 2, \xi_t(x) = 2) \\ &= -\delta_2 u + \beta_2 u(1 - u) \end{aligned}$$

where the first equality is true in general and the second follows from our assumptions and the fact that $\beta_2 = |\mathcal{N}| \lambda_2$. Dropping the $-\beta_2 u^2$ term

$$\frac{du}{dt} \le (\beta_2 - \delta_2) u$$

so if $\delta_2 > \beta_2$ all solutions tend to 0 exponentially fast. If $\delta_2 < \beta_2$ and we let $u^* = (\beta_2 - \delta_2)/\beta_2$ then

$$-\delta_2 u + \beta_2 u(1 - u) \begin{cases} > 0 & \text{for } 0 < u < u^* \\ < 0 & \text{for } u > u^* \end{cases}$$

so if $u(0) > 0$, $u(t) \to u^*$ as $t \to \infty$.

Applying the reasoning that led to (1.1) to $v(t) = P(\xi_t(x) = 1)$ we see that

$$(1.2) \qquad \begin{aligned} \frac{dv}{dt} &= -\delta_1 P(\xi_t(x) = 1) - \sum_{y \sim x} \lambda_2 P(\xi_t(x) = 1, \xi_t(y) = 2) \\ &\quad + \sum_{y \sim x} \lambda_1 P(\xi_t(x) = 0, \xi_t(y) = 1) \\ &= -\delta_1 v - \beta_2 v u + \beta_1 (1 - u - v) v \end{aligned}$$

where again the first equality is true in general and the second follows from our assumptions and the fact that $\beta_i = |\mathcal{N}|\lambda_i$. To analyze (1.2), we note that if the 2's are in equilibrium and the density of 1's is very small, then

$$u = (\beta_2 - \delta_2)/\beta_2 \qquad (1 - u - v) \approx \delta_2/\beta_2$$

$$\text{1's are born at rate } \approx \beta_1 \cdot \frac{\delta_2}{\beta_2} \cdot v$$

$$\text{1's die at rate } \approx \left(\delta_1 + \beta_2 \cdot \frac{\beta_2 - \delta_2}{\beta_2} \right) v$$

So if $(\star)$ holds a small density of 1's will grow in time, while if we reverse the inequality in $(\star)$ and use $(1 - u - v) \leq (\beta_2 - \delta_2)/\beta_2$ then the birth rate always exceeds the death rate and $v(t) \to 0$.

The practice of calculating how densities evolve when we suppose that adjacent sites are independent is called *mean field theory*. Theorems 5C and 5D are one instance of the general principle that when the range of interaction is large mean field calculations are almost correct. A second method of making mean field calculations correct, which leads to connections with nonlinear partial differential equations, is to introduce particle motion at a fast rate.

Example 1.6. Predator prey systems. In this model we think of 0 = vacant, 1 = occupied by a fish, and 2 = occupied by a shark and we have the following flip rates

$$c_1(x, \xi) = \beta_1 n_1(x, \xi)/2d \quad \text{if } \xi(x) = 0$$
$$c_2(x, \xi) = \beta_2 n_2(x, \xi)/2d \quad \text{if } \xi(x) = 1$$
$$c_0(x, \xi) = \begin{cases} \delta_1 & \text{if } \xi(x) = 1 \\ \delta_2 + (\gamma n_2(x, \xi)/2d) & \text{if } \xi(x) = 2 \end{cases}$$

In words, fish die at rate δ_1 and are born at vacant sites at a rate proportional to the number of fish at neighboring sites. So in the absence of sharks, the fish are a contact process.

Sharks die of natural causes at rate δ_2 and kill a neighboring shark at rate $\gamma/2d$. The birth rate for sharks may look a little strange at first: fish turn into sharks at rate proportional to the number of shark neighbors. This is not what happens in the ocean but it does capture an essential feature of the interaction: when the density of fish is too low then the sharks die faster than they give birth. A second justification of this mechanism is that, as we will see in Section 9, in a suitable limit we get standard predator-prey equations. Here $n_i(x, \xi) = |\{z \in \mathcal{N} : \xi(x + z) = i\}|$ as usual, but for reasons that will become clear in a moment we take $S = \epsilon \mathbf{Z}^d$ and $\mathcal{N} = \{z : |z| = \epsilon\}$ the nearest neighbors. We use a small lattice so that we can introduce *stirring* at a fast rate, i.e., for each $x, y \in \epsilon \mathbf{Z}^d$ with $|x - y| = \epsilon$ we exchange the values at x and y at rate ϵ^{-2}. That is, we change the configuration from ξ to $\xi^{x,y}$ defined by

$$\xi^{x,y}(x) = \xi(y), \quad \xi^{x,y}(y) = \xi(x), \quad \xi^{x,y}(z) = \xi(z) \text{ if } z \neq x, y$$

The combination of the space scale of ϵ and the time scale of ϵ^{-2} means that the individual values will perform Brownian motions in the limit $\epsilon \to 0$. The fast stirring keeps the states of neighboring sites independent, so using mean field reasoning leads to the following result due to DeMasi, Ferrari and Lebowitz (1986).

Theorem 6A. Suppose $\xi_0^\epsilon(x)$, $x \in \epsilon\mathbf{Z}^d$, are independent and let $u_i^\epsilon(t, x) = P(\xi_t(x) = i)$. If $u_i^\epsilon(0, x) = g_i(x)$ is continuous then as $\epsilon \to 0$, $u_i^\epsilon(t, x)$ converges to $u_i(t, x)$ the bounded solution of

(1.3)
$$\frac{\partial u_1}{\partial t} = \Delta u_1 + \beta_1 u_1(1 - u_1 - u_2) - \beta_2 u_1 u_2 - \delta_1 u_1$$
$$\frac{\partial u_2}{\partial t} = \Delta u_2 + \beta_2 u_1 u_2 - \delta_2 u_2 - \gamma u_2^2$$

with $u_i(0, x) = g_i(x)$.

Here the Δu_i terms reflect the fact that in the limit the individual values are performing Brownian motions run at rate 2. The other terms can be seen by using the reasoning that led to (1.1) and (1.2).

If we suppose that the initial functions $g_i(x)$ are constant then this is true at later times $u_i(t, x) = v_i(t)$ and the v_i satisfy

(1.4)
$$\frac{\partial v_1}{\partial t} = v_1((\beta_1 - \delta_1) - \beta_1 v_1 - (\beta_1 + \beta_2)v_2)$$
$$\frac{\partial v_2}{\partial t} = v_2(-\delta_2 + \beta_2 v_1 - \gamma v_2)$$

Here we have rearranged the right hand side to show that it is the standard predator-prey equations with limited growth. (See for example Hirsch and Smale (1974) p. 263.) To determine the conditions for coexistence, we start by finding the fixed points of the dynamical systems, i.e., points (ρ_1, ρ_2) so that $v_i(t) \equiv \rho_i$ is a solution of (1.4). There are three

(i) $\rho_1 = \rho_2 = 0$. No sharks or fish, the trivial equilibrium.

(ii) We have a solution with $\rho_2 = 0$ and $\rho_1 = (\beta_1 - \delta_1)/\beta_1$ if $\beta_1 > \beta_2$. This forumla is the same as the one in the last example because in the absence of sharks, fish are a contact process.

(iii) There is a fixed point with $\rho_i = \sigma_i > 0$ if and only if

(1.5)
$$\frac{\beta_1 - \delta_1}{\beta_1} > \frac{\delta_2}{\beta_2}$$

(which implies $\beta_1 > \delta_1$). We do not have an intuitive explanation for the last condition. It is simply what results when we solve the two equations in two unknowns.

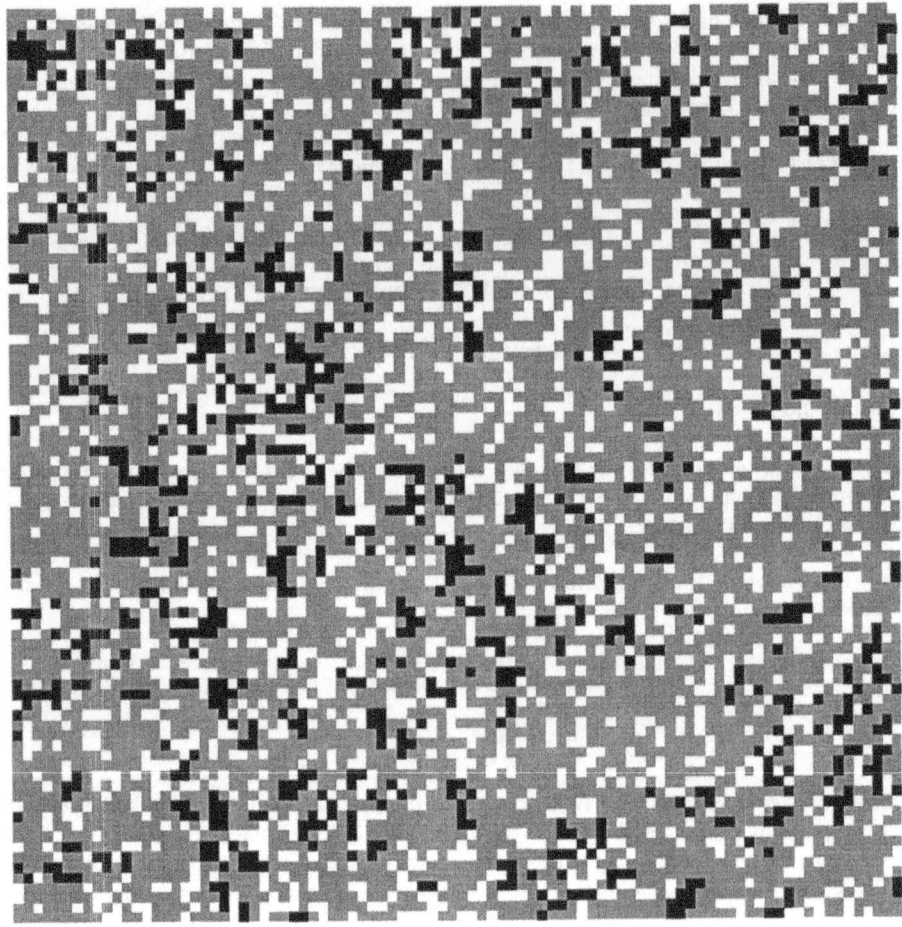

Figure 1.6. Predator prey model $\beta_1 = \beta_2 = 3$, $\delta_1 = \delta_2 = 1$, $\gamma = 1$

By exploiting the connection between the particle system and the partial differential equation given in Theorem 6A, we can prove

Theorem 6B. If (1.5) holds then for small ϵ coexistence occurs.

It would be nice to prove coexistence results without fast stirring. Figure 1.6 shows a simulation of the system on $\{0, 1, \ldots, 79\}^2$ at time 50 with $\beta_1 = \beta_2 = 3$, $\delta_1 = \delta_2 = 1$, $\gamma = 1$ and no stirring. Again sites in state 1 are gray; those in state 2 are black.

Example 1.7. Epidemic model. In this example, we think of $\mathbf{Z}^2$ as representing an array of houses each of which is occupied by one individual who can be (0) susceptible = healthy but capable of getting the disease, (1) infected with the disease, or (2) immune to further infection. The flip rates are

$$c_1(x, \xi) = \lambda n_1(x, \xi) \quad \text{if } \xi(x) = 0$$
$$c_2(x, \xi) = \delta \quad \text{if } \xi(x) = 1$$
$$c_0(x, \xi) = \alpha \quad \text{if } \xi(x) = 2$$

As usual, the rates we did not mention are 0. In words, a susceptible individual gets infected at a rate proportional to the number of infected neighbors. Infected individuals become removed at rate δ. Here $1/\delta$ is the mean duration of the disease and to obtain the Markov property we have assumed that the duration of the disease has an exponential distribution. If we want to model the short term behavior of a measles or flu epiemic then we set $\alpha = 0$ since recovered individuals are immune to the disease. If we want to examine longer time properties then immune individuals will die (or move out of town) and new susceptibles will be born (or move into town) so to keep a fixed population size of one individual per site, we combine the two transitions into one.

To describe the conditions for coexistence we begin with case $\alpha = 0$ and consider the behavior of the model starting from one infected individual at 0 in the midst of an otherwise susceptible population. Let $\eta_t = \{x : \xi_t(x) = 1\}$ be the set of the infected individuals at time t and let $\tau = \inf\{t : \eta_t = \emptyset\}$. We will have $\eta_t = \emptyset$ for all $t > \tau$ so we say the infection *dies out* at time τ. Let $\delta_c = \inf\{\delta : P(\tau = \infty) = 0\}$. The faster people recover the harder it is for the epidemic to propagate so we have $P(\tau = \infty) = 0$ for all $\delta > \delta_c$.

If we restrict our attention to the nearest neighbor case, then results of Cox and Durrett (1988) describe the asymptotic behavior of the epidemic when $\delta < \delta_c$ and $\tau = \infty$. Building on those results Durrett and Neuhauser (1991) have shown

Theorem 7. Suppose $d = 2$ and $\mathcal{N} = \{x : |x| = 1\}$. If $\delta < \delta_c$ and $\alpha > 0$ then coexistence occurs.

Zhang has generalized the results of Cox and Durrett (1988) to finite range interactions. Presumably one can also prove the result of Durrett and Neuhauser (1991) in that level of generality but no one has had the courage to try to write out all the details.

Closely related to the epidemic model is

Example 1.8. Greenberg Hastings Model. In this model, we think of having a neuron at each $x \in \mathbf{Z}^d$ that is connected to each of its neighbors. The states of each neuron are $F = \{0, 1, \ldots, \kappa - 1\}$ where 1 is excited, $2, \ldots, \kappa - 1$ are a sequence of recovery states, and 0 indicates a fully rested neuron that is capable of being excited. These interpretations motivate the following flip rates

$$c_1(x, \xi) = 1 \quad \text{if } \xi(x) = 0 \text{ and } n_i(x, \xi) \geq \theta$$
$$c_i(x, \xi) = 1 \quad \text{if } i \neq 1 \text{ and } \xi(x) = i - 1$$

Here arithmetic is done modulo κ so $0 - 1 = \kappa - 1$. The second rule says that once excited, the neuron progresses through the recovery states at rate 1 until it is fully rested; the first that a rested neuron becomes excited at rate 1 if the number of its neighbors that are excited is at least the threshold θ. The next result, due to Durrett (1992), gives a regime in which this model has (somewhat boring) stationary distributions.

Theorem 8A. Let $\epsilon > 0$ and suppose $\theta \leq (1 - \epsilon)|\mathcal{N}|/2\kappa$. If r is large then there is a stationary measure close to the uniform product measure.

Here the *uniform product measure* is the one in which the coordinates $\xi(x)$ are independent and $P(\xi(x) = i) = 1/\kappa$. Based on the analogy with the epidemic model where if $\delta < \delta_c$ there is a coexistence for any $\alpha > 0$, we expect that

Conjecture 8B. There is a constant $a > 0$ so that if $\theta \leq a|\mathcal{N}|$ then coexistence occurs for any κ.

Computer simulations indicate that in this regime the excitation sustains itself by producing moving fronts. See Figure 1.7 for a simulation of the system with $\mathcal{N} = \{x : \|x\|_\infty \leq 2\}$, threshold $\theta = 3$, and $\kappa = 8$. Excited states are black, rested sites are white, recovering sites are appropriate shades of gray.

The analogue of Conjecture 8B has been proved by Durrett and Griffeath (1993) for the Greenberg Hastings cellular automaton in which $\xi_{n+1}(x) = \xi_n(x) + 1$ if $\xi_n(x) > 0$ or $\xi_n(x) = 0$ and $n_i(x, \xi_n) \geq \theta$; $\xi_{n+1}(x) = \xi_n(x)$ otherwise. See Figure 1.8 for a simulation of the cellular automaton with the same color scheme and parameters: $\mathcal{N} = \{x : \|x\|_\infty \leq 2\}$, threshold $\theta = 3$, and $\kappa = 8$ run until it has become periodic with period 8. For more on the cellular automaton consult Fisch, Gravner and Griffeath (1991), (1992), (1993), and Gravner and Griffeath (1993).

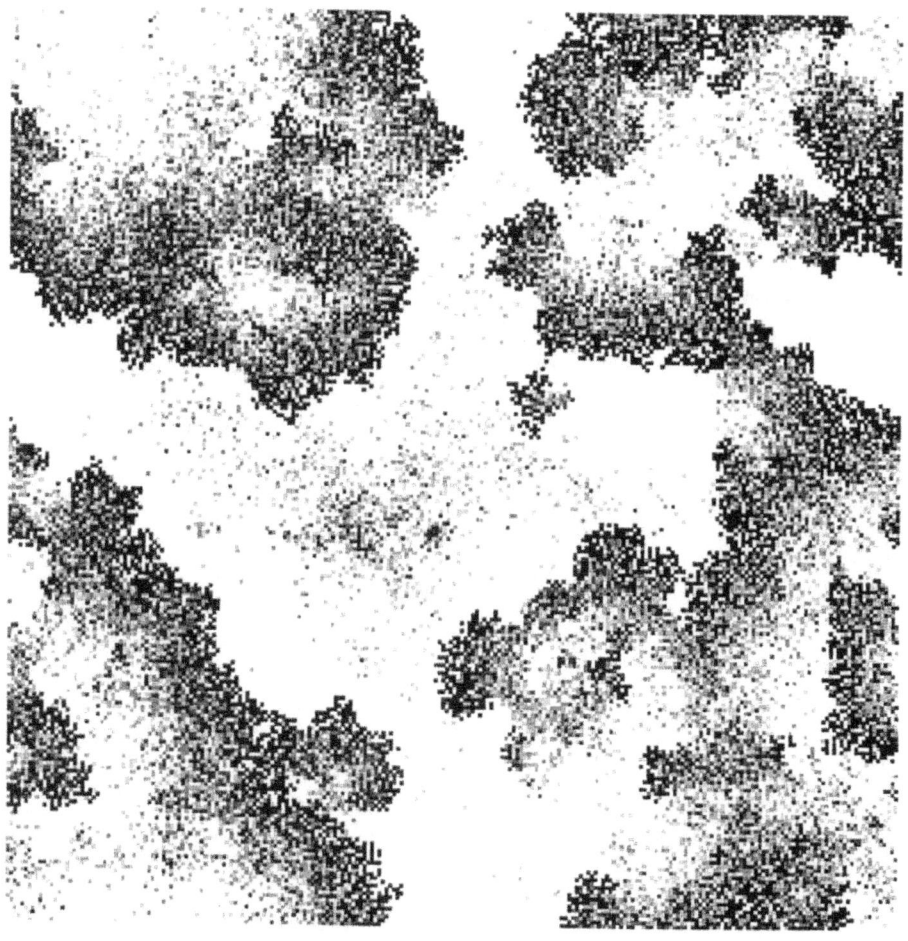

Figure 1.7. Greenberg Hastings model. $\mathcal{N} = \{x : \|x\|_\infty \leq 2\}$, $\theta = 3$, $\kappa = 8$

Figure 1.8. Greenberg Hastings cellular automaton. $\mathcal{N} = \{x : \|x\|_\infty \leq 2\}$, $\theta = 3$, $\kappa = 8$

2. Construction, Basic Properties

To construct an interacting particle system from given translation invariant finite flip rates

$$c_i(x, \xi) = g_i(\xi(x + z_0), \xi(x + z_1), \ldots, \xi(x + z_k))$$

based on a neighborhood set $\mathcal{N} = \{z_0, z_1, \ldots, z_n\}$ we can, by changing the time scale, assume that $c_i(x, \xi) \leq 1$. For each $x \in \mathbf{Z}^d$ and $i \in F$, let $\{T_n^{x,i} : n \geq 1\}$ be the arrival times of independent rate 1 Poisson processes (i.e., if we set $T_0^{x,i} = 0$ then the increments $T_n^{x,i} - T_{n-1}^{x,i}$ are independent and have an exponential distribution with mean 1) and let $U_n^{x,i}$ be independent and uniform on (0,1). At time $t = T_n^{x,i}$ site x will flip to state i if $U_n^{x,i} < c_i(x, \xi_{t-})$ and stay unchanged otherwise. To see that this recipe produces the desired flip rates recall the *thinning property* of the Poisson process: if we keep the points from $\{T_n^{x,i} : n \geq 1\}$ that have $U_n^{x,i} < p$ then the result is a Poisson process with rate p.

Since there are infinitely many Poisson processes, and hence no first arrival, we have to show that we can use our recipe to compute the time evolution. To do this, we use an argument of Harris (1972). Let t_0 be a small positive number to be chosen later. We draw an unoriented arc between x and y if $y - x \in \mathcal{N}$ and for some i, $T_1^{x,i} < t_0$. The presence of an arc between x and y indicates that a Poisson arrival has caused x to look at y to see if it should flip or caused y to look at x. Conversely, if there is no arc between x and y then neither site has looked at the other. The last observation implies that the sites in two different components of the resulting random graph have not influenced each other by time t_0 and hence their evolutions can be computed separately. To finish the construction then it suffices to show

(2.1) Theorem. If t_0 is small enough then with probability one, all the connected components of our random graph are finite.

For then in each component there is a first flip and we can compute the effects of the changes sequentially. This allows us to construct the process up to time t_0 but t_0 is independent of the initial configuration, so iterating we can construct the process for all time.

PROOF OF (2.1): Let $\mathcal{N}^* = \{z_1, \ldots, z_k, -z_1, \ldots, -z_k\}$ be the set of possible displacements along edges of the graph. (In this section alone, we will allow $\mathcal{N}$ to be a general finite set not just $\{x : \|x\|_p \leq r\}$.) We say that $y_0, y_1, \ldots y_n$ is a *path* of length n if $y_m - y_{m-1} \in \mathcal{N}^*$ when $0 < m \leq n$. We call a path *self-avoiding* if $y_i \neq y_j$ when $0 \leq i < j \leq n$. Let $R = \max\{|z_i| : z_i \in \mathcal{N}\}$. (Here $|z| = \|z\|_2$.) We claim that

(a) If 0 is connected to some point with $|z| > M$ then there is a self-avoiding path of length $\geq M/R$ starting at 0.

To see this note that if there is a path from 0 to z, then by removing loops we can make it self-avoiding. Since each step along the path moves us a distance $\leq R$, there must be at least $|z|/R$ such steps. The next ingredient in the proof is

(b) If x, y, z, w are distinct, the presence of edges from x to y and from z to w are independent events.

To see this note that the presence of an edge from x to y is determined by the Poisson processes $T_n^{x,i}$ and $T_n^{y,i}$ with $i \in F$. From (b) it is easy to see

(c) Let $N = |\mathcal{N}^*|$ and $\kappa = |F|$. The probability of a self avoiding path of length $2n - 1$ starting at a given point x is at most

$$N^{2n-1}(1 - e^{-2\kappa t_0})^n$$

The first factor is the number of paths of length $2n - 1$ and hence an upper bound on the number of self-avoiding paths. To see the second factor note that the presence of the edges $(z_0, z_1), (z_2, z_3), \ldots (z_{2n-2}, z_{2n-1})$ are independent events that have probability $1 - e^{-2\kappa t_0}$ since the probability of no arrival by time t_0 in one of the 2κ Poisson processes $T_n^{x,i}$ and $T_n^{y,i}$ is $e^{-2\kappa t_0}$.

If we pick t_0 small enough then $N^2(1 - e^{-2\kappa t_0}) \leq 1/2$, so the probability of a self-avoiding path of length $2n - 1$ decreases to 0 exponentially fast, and it follows from (a) that with probability 1 the cluster containing any given point x is finite. □

An immediate consequence of the construction is

(2.2) **Corollary.** If ξ_0 is translation invariant then ξ_t is.

PROOF: The family of Poisson processes is translation invariant, so if the initial state is, then so is the result of our computation. □

It should also be clear from the construction that ξ_t is a Markov process, i.e., if we know the state at time s, information about ξ_r for $r < s$ is irrelevant for computing the evolution for $t > s$. Being a Markov process there is an associated family of operators defined by

$$T_t f(\xi) = E_\xi f(\xi_t)$$

where E_ξ denotes the expected value starting from $\xi_0 = \xi$. The Markov property of ξ_t implies that the T_t form a *semigroup*. That is, $T_s T_t = T_{s+t}$. If you are not familiar with semi-groups don't worry. We will only use the most basic results that can be found in Chapter 1 of Dynkin (1965) or in Chapter ? of Revuz and Yor (1991), and we will only use those facts in this section. The first thing we want to prove is

(2.3) **Corollary.** T_t is a *Feller semigroup*, i.e., if f is continuous with respect to the product topology on F^S then $T_t f$ is continuous.

PROOF: Note that our construction defines on the same probability space the process starting from any initial configuration. If $t \leq t_0$ then proof of (2.1) shows that up to time t_0, $\mathbf{Z}^d$ breaks up into a collection of finite non-interacting islands. From the last fact it follows easily that if $\xi_0^n \to \xi_0$, (which means that for each fixed x, $\xi_0^n(x) \to \xi_0(x)$) then $\xi_t^n \to \xi_t$ almost surely. If f is continuous it follows that $f(\xi_t^n) \to f(\xi_t)$ almost surely. Since F^S is compact in the product topology, any continuous function is necessarily bounded, and it follows from the bounded convergence theorem that $Ef(\xi_t^n) \to Ef(\xi_t)$. This proves the result for $t \leq t_0$. Using the semigroup property $T_{t+s} = T_s T_t$, it follows that the result holds for $t \leq 2t_0$, $t \leq 3t_0$, and hence for all t. □

Our next step is to compute the generator of the semigroup. Let $\xi^{x,i}$ denote the configuration ξ flipped to i at x. That is,

$$\xi^{x,i}(x) = i \qquad \xi^{x,i}(y) = \xi(y) \quad \text{otherwise}$$

Suppose $f(\xi)$ only depends on the values of finitely many coordinates and let

$$Lf = \sum_{x \in Z^d, i \in F} c_i(x, \xi) \left(f(\xi^{x,i}) - f(\xi) \right)$$

The sum converges since only finitely many terms are nonzero. Our next result says that L is the generator of T_t.

(2.4)
$$\frac{d}{dt} T_t f(\xi) \Big|_{t=0} = Lf(\xi)$$

If you have seen the generator of a Markov process with a discrete state space the formula should not be surprising. The proof of (2.4) is much like the proof for that case so we will only give a quick sketch.

PROOF: Suppose f only depends on the values of ξ in $[-L, L]^d$ and recall we have defined $R = \max\{|z_i| : z_i \in N\}$. If t is small then with high probability there is at most one site $x \in [-L - 2R, L + 2R]^d$ and one value of $i \in F$ with $T^{x,i} < t$. By considering the various possible values of x and i and noting that the probability that $\xi_0 = \xi$ changes to $\xi^{x,i}$ is $\sim tc_i(x, \xi)$, the result follows easily. □

For the rest of this section, we will restrict our attention to the case $F = \{0, 1\}$, in which case we think of $1 =$ occupied by a particle and $0 =$ vacant. Since we think of 1's are particles we call $c_1(x, \xi)$ the *birth rates* and call $c_0(x, \xi)$ the *death rates*. We say that the birth rates $c_1(x, \xi)$ are *increasing* if

$$\xi(y) \leq \zeta(y) \text{ for all } y \neq x \text{ and } \xi(x) = \zeta(x) = 0 \text{ implies } c_1(x, \xi) \leq c_1(x, \zeta)$$

We say that *death rates* $c_0(x, \xi)$ are *decreasing* if

$$\xi(y) \leq \zeta(y) \text{ for all } y \neq x \text{ and } \xi(x) = \zeta(x) = 1 \text{ implies } c_1(x, \xi) \geq c_1(x, \zeta)$$

A process with increasing birth rates and decreasing death rates is said to be *attractive*. The last term comes from analogies with the Ising model in statistical mechanics. This assumption is not very attractive for biological systems since there the death rate usually increases due to crowding, but the attractive property is what we need to prove the following useful result.

(2.5) **Theorem.** For an attractive process, if we are given initial configurations with $\xi_0(x) \leq \zeta_0(x)$ for all x then the processes defined by our construction have $\xi_t(x) \leq \zeta_t(x)$ for all x and t.

PROOF: Intuitively this is true since each flip preserves the inequality. To check this suppose that $\xi_{s-}(y) = 0$ and a birth event $T_n^{y,1}$ occurs at time s. If $\zeta_{s-}(y) = 1$ then $\zeta_s(y) = 1$ and the inequality will certainly hold after the flip. If $\zeta_{s-}(y) = 0$ and the inequality holds before the flip, then since our birth rates are increasing $c(y, \xi_{s-}) \leq c(y, \zeta_{s-})$. By considering the possible values of $U_n^{y,i}$ se see that in all cases the inequality holds after the flip.

value of $U_n^{y,i}$	change in ξ	change in ζ
$[0, c(y, \xi_{s-}))$	flips to 1	flips to 1
$[c(y, \xi_{s-}), c(y, \zeta_{s-}))$	stays 0	flips to 1
$[c(y, \zeta_{s-}), 1)$	stays 0	stays 0

A similar argument applies if $\xi_{s-}(y) = 1$ and a death event $T_n^{y,0}$ occurs at time s.

To turn the intuitive argument in the last paragraph into a proof, suppose that the inequality fails at some point x at some time $t \leq t_0$. Let C_x be the connected component containing x for the random graph defined in the proof of (3.1), and let $s > 0$ be the first time the property fails at some point $y \in C_x$. By the definition of s the inequality holds on C_x before time s. Since C_x contains all the neighbors of any site in C_x that flips by time t_0 it follow from the argument in the last paragraph that the inequality will hold until the next flip after time s. Since C_x is a finite set, the next flip will occur at a time $> s$, contradicting the defintion of s and showing that the inequality must hold up to time t_0. Iterating the last conclusion we see that the result holds for all time. □

To explain our interest in (2.2), (2.4), and (2.5) we will now prove that

(2.6) **Theorem.** If $\lambda|\mathcal{N}| < \delta$ then the contact process has no nontrivial stationary distribution.

PROOF: Consider the contact process starting from all sites occupied, i.e., suppose $\xi_0^1(x) = 1$ for all x. It follows from (2.2) that $P(\xi_t^1(x) = 1)$ is independent of x, so writing $y \sim x$ for "y is a neighbor of x" and using $\frac{d}{dt}T_t f = T_t L f$ we have

$$\frac{d}{dt} P(\xi_t^1(x) = 1) = -\delta P(\xi_t^1(x) = 1) + \sum_{y \sim x} \lambda P(\xi_t^1(x) = 0, \xi_t^1(y) = 1)$$

$$\leq -\delta P(\xi_t^1(x) = 1) + \lambda|\mathcal{N}| P(\xi_t^1(y) = 1)$$

If $\lambda|\mathcal{N}| < \delta$ then the last inequality implies that $P(\xi_t^1(x) = 1) \to 0$ as $t \to \infty$. Now any initial configuration has $\xi_0(x) \leq 1 = \xi_0^1(x)$ for all x, so by (2.5), we have $\xi_t(x) \leq \xi_t^1(x)$ for all t and x and it follows that $P(\xi_t(x) = 1) \to 0$ for any initial configuration. If we pick ξ_0 to have a stationary distribution then $P(\xi_t(x) = 1)$ is independent of t, so the last conclusion implies this probability is 0 and the result follows. □

The last argument shows that if we start an attractive process with all sites occupied and find $P(\xi_t^1(x) = 1) \to 0$ then there is no nontrivial stationary distribution. Our next result proves the converse. Recall that $\Rightarrow$ denotes weak convergence, which in this setting is just convergence of finite dimensional distribution.

(2.7) Theorem. As $t \to \infty$, $\xi_t^1 \Rightarrow \xi_\infty^1$. The limit is a stationary distribution which is stochastically larger than any other stationary distribution and called the *upper invariant measure*.

PROOF: The key to the proof is the following observation:

(2.8) Lemma. For any set $A \subset \mathbf{Z}^d$, $t \to P(\xi_t^1(x) = 0$ for all $x \in A)$ is increasing.

PROOF: Let $\zeta_0 = \xi_s^1$. Clearly, $\xi_0^1(x) \geq \zeta_0(x)$ for all x so (2.5) implies that for all t and x, $\xi_t^1(x) \geq \zeta_t(x)$. Since ζ_t has the same distribution as ξ_{s+t}^1 it follows that

$$P(\xi_t^1(x) = 0 \text{ for all } x \in A) \leq P(\xi_{s+t}^1(x) = 0 \text{ for all } x \in A) \qquad \square$$

Let $\phi(A) = P(\xi(x) = 0$ for all $x \in A)$ and $B = \{x_1, \ldots, x_m\}$ Using the inclusion exclusion formula on the events $E_i = \{\xi(x) = 0\}$ on $A \cup \{x_i\}$, we can express any finite dimensional distribution in terms of the $\phi(C)$.

$$1 - P(\xi(x) = 0 \text{ for all } x \in A, \xi(x) = 1 \text{ for all } x \in B) = P(\cup_{i=1}^m E_i)$$

$$= \sum_{i=1}^m \phi(A \cup \{x_i\}) - \sum_{i<j} \phi(A \cup \{x_i, x_j\}) + \ldots + (-1)^{m+1} \phi(A \cup B)$$

So (2.8) implies convergence of all finite dimensional distributions. $\qquad \square$

The fact that ξ_∞^1 is a stationary distribution follows from a general result.

(2.9) Lemma. Suppose the Markov process X has a Feller semigroup and $X_t \Rightarrow X_\infty$ then (the distribution of) X_∞ is a stationary distribution.

PROOF: Recall that if X_0 has distribution μ then the probability measure μT_t defined by

$$\int (\mu T_t)(dx) f(x) = \int \mu(dx) T_t f(x) = \int \mu(dx) E_x f(X_t)$$

for all bounded continuous functions f gives the distribution of X_t when X_0 has distribution μ. The key to the proof of (2.9) is the following general fact:

(2.10) If T_t is a Feller semigroup and $\mu_s \Rightarrow \mu$ then $\mu_s T_t \Rightarrow \mu T_t$.

To prove (2.10) we note that $T_t f$ is bounded and continuous

$$\lim_{s \to \infty} \int (\mu_s T_t)(dx) f(x) = \lim_{s \to \infty} \int \mu_s(dx) T_t f(x)$$

$$= \int \mu(dx) T_t f(x) = \int (\mu T_t)(dx) f(x)$$

where the second inequality follows from the fact that $T_t f$ is continuous and $\mu_s \Rightarrow \mu$. To prove (2.9) now, let μ_s be the distribution of X_s and note that the Markov property implies $\mu_s T_t = \mu_{s+t}$. The right hand side converges to μ, and by (2.10) the left hand side converges to μT_t, so $\mu T_t = \mu$, i.e., μ is a stationary distribution. □

Finally we have to explain and show the claim "ξ_∞^1 is stochastically larger than any other stationary distribution π." By *stochastically larger* we mean that if f is any increasing function which depends on only finitely many cooordinates then

$$(2.11) \qquad Ef(\xi_\infty^1) \geq \int f(\xi)d\pi(\xi)$$

Here f is *increasing* means that if $\xi(x) \leq \zeta(x)$ for all x then $f(\xi(x)) \leq f(\zeta(x))$. To prove the claim let ζ_0 have distribution π. Clearly, $\xi_0^1(x) \geq \zeta_0(x)$ for all x so (3.5) implies that $\xi_t(x) \geq \zeta_t(x)$ for all t and x. Now if f is increasing

$$Ef(\xi_t) \geq Ef(\zeta_t) = \int f(\xi)d\pi(\xi)$$

since π is a stationary distribution. If f depends on only finitely many coordinates then it is continuous and

$$Ef(\xi_t^1) \to Ef(\xi_\infty^1)$$

Combining the last two conclusions, proves our claim and completes the proof of (3.7). □

(2.12) **Remark.** A result of Holley implies that since ξ_∞^1 is stochastically larger than π, we can define random variables ξ and ζ with these distributions on the same probability space so that $\xi(x) \geq \zeta(x)$.

Later we will need a variation of (2.9). The next result and (3.15) are not needed until Section 5, so I suggest that you wait until later to read the rest of this section.

(2.13) **Theorem.** Suppose the Markov process X has a compact state space Λ and a Feller semigroup T_t. Let μ_t be the distribution of X_t and ν_t the Cesaro average defined by

$$\nu_t(A) = \frac{1}{t}\int_0^t \mu_s(A)$$

If $t_k \to \infty$ and $\nu_{t_k} \Rightarrow \nu$ then ν is a stationary distribution.

(2.14) **Corollary.** Since the set of probability measures on Λ is compact in the weak topology, this implies in particular that stationary distributions exist.

PROOF: Since $\mu_s T_r = \mu_{s+r}$ we have

$$\nu_{t_k} T_r = \frac{1}{t_k}\int_0^{t_k} \mu_s T_r ds = \frac{1}{t_k}\int_r^{r+t_k} \mu_s ds$$

$$= \nu_{t_k} + \frac{1}{t_k}\int_{t_k}^{r+t_k} \mu_s ds - \frac{1}{t_k}\int_0^r \mu_s ds$$

The two error terms on the right hand side have each total mass r/t_k and hence converge weakly to 0. Since $\nu_{t_k} \Rightarrow \nu$ it follows that $\nu_{t_k} T_r \Rightarrow \nu$. On the other hand it follows from (3.10) that $\nu_{t_k} T_r \Rightarrow \nu T_r$ so we have $\nu T_r = \nu$ as desired. $\qquad \square$

In Section 5, we will also need the following result:

(2.15) **Theorem.** The upper invariant measure ξ_∞^1 is spatially ergodic.

PROOF: We begin with the observation that

(2.16) for each t, ξ_t^1 is spatially ergodic.

To prove (2.16) we let $V^x = (\{T_n^{x,i}, n \geq 1\}, \{U_n^{x,i}, n \geq 1\}, i = 0, \ldots, \kappa - 1)$. $\{V^x, x \in \mathbf{Z}^d\}$ are i.i.d. and $\xi_t(x)$ is a function of the V_x so the result follows from a generalization of (1.3) in Chapter 6 of Durrett (1992). In words, functions of ergodic sequences are ergodic.

To let $t \to \infty$, we note that the proof of (2.8) shows ξ_t^1 is stochastically larger than ξ_∞^1 so (2.12) implies that we can construct the two processes on the same space so that $\xi_t^1(x) \geq \xi_\infty^1(x)$ for all x. Let f be an increasing function that depends on only finitely many coordinates. The ergodic theorem implies that as $L \to \infty$

$$\frac{1}{(2L+1)^d} \sum_{x : \|x\|_\infty \leq L} \xi_t^1(x) \to E f(\xi_t)$$

$$\frac{1}{(2L+1)^d} \sum_{x : \|x\|_\infty \leq L} \xi_\infty^1(x) \to E(f(\xi_\infty)|\mathcal{I})$$

The last result and our comparison imply that $E(f(\xi_\infty)|\mathcal{I}) \leq E f(\xi_t)$ where $\mathcal{I}$ is the σ-field of shift invariant events. Letting $t \to \infty$ we have $E(f(\xi_\infty)|\mathcal{I}) \leq E f(\xi_\infty)$ and since the left hand side has expected value $E f(\xi_\infty)$, it follows that

(2.17) $$E(f(\xi_\infty)|\mathcal{I}) = E f(\xi_\infty) \qquad \text{a.s.}$$

At this point we have shown that (2.17) holds for increasing functions that depends on only finitely many coordinates. Now every function on $\{0,1\}^k$ is a difference of two increasing functions so (2.17) holds for any function of finitely many coordinates. Taking limits and using the inequality

$$E|E(X - Y|\mathcal{I})| \leq E(|X - Y||\mathcal{I}) = E|X - Y|$$

shows that (2.17) holds for all bounded f so $\mathcal{I}$ is trivial. $\qquad \square$

3. Percolation Substructures, Duality

In this section we introduce a variation of the construction used in Section 2, due to Harris (1976) and Griffeath (1979), which applies to a special class of models with state space $\{0,1\}^S$ and leads to a "duality relationship." For these purposes it is convenient to write our systems as set valued processes in which the state at time t is the set of sites occupied by 1's. We begin with

Example 3.1. The basic contact process. We let $\mathcal{N}$ be a finite set of neighbors of 0, say that y is a neighbor of x if $y - x \in \mathcal{N}$, and formulate the dynamics as follows:

(i) Particles die at rate 1.

(ii) A particle is born at a vacant site x at rate λ times the number of occupied neighbors.

To construct the process we introduce independent Poisson processes $\{U_n^x, n \geq 1\}$ with rate 1 and $\{T_n^{x,y}, n \geq 1\}$ with rate λ for each $x, y \in \mathbf{Z}^d$ with $y - x \in \mathcal{N}$. At the space time points (x, U_n^x) we write a δ to indicate that a death will occur if x is occupied, and we draw an arrow from $(y, T_n^{x,y})$ to $(x, T_n^{x,y})$ to indicate that if y is occupied then there will be a birth from y to x.

Given the Poisson processes and forgetting about the special marks, we could construct the process using the algorithm described in the last section. We introduce the special marks to make contact with percolation: we imagine fluid entering the bottom of the picture at the points in ξ_0 and flowing up the structure. The δ's are dams, the arrows are pipes that allow the fluid to flow in the direction of the arrow, and ξ_t is the set of sites that are wet at time t.

An example of the *percolation substructure* and the corresponding realization of ξ_t starting from $\xi_0 = \{0, 1\}$ is given in Figure 3.1. The thick lines indicate the sites that are occupied. To be able to define the dual process, we need an explicit recipe for constructing ξ_t from the picture. We say that there is a *path from* $(x, 0)$ *to* (y, t) if there is a sequence of times $s_0 = 0 < s_1 < s_2 < s_n < s_{n+1} = t$ and spatial locations $x_0 = x, x_1, \ldots, x_n = y$ so that

(i) for $i = 1, 2, \ldots, n$ there is an arrow from x_{i-1} to x_i at time s_i

(ii) the vertical segments $\{x_i\} \times (s_i, s_{i+1})$, $i = 0, 1, \ldots n$ do not contain any δ's.

(Exercise: Find a path from $(2, 0)$ to $(3, t)$ in Figure 3.1.) Intutitively the arrows are births that will occur if there are no δ's in the intervals in (ii), so to define the process starting from $\xi_0^A = A$ we let

(3.1) $\xi_t^A = \{y : \text{for some } x \in A \text{ there is a path from } (x, 0) \text{ to } (y, t)\}$

It should be clear from the definitions that ξ_t^A is the contact process with one small modification: because of the open intervals in (ii) and the strict inequality in $s_n < s_{n+1} = t$, the process we have constructed is left continuous. For example, if there is a death at x at time t, the particle will not be dead at time t but it will be dead at time $t + \epsilon$ when ϵ

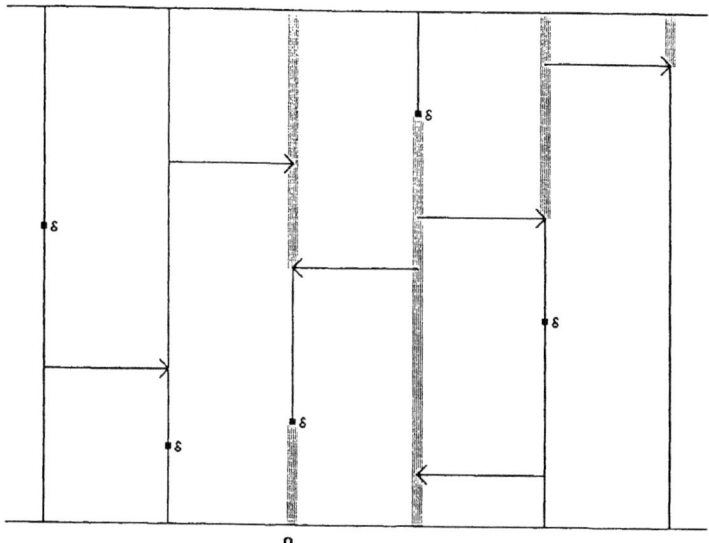

Figure 3.1. Contact process

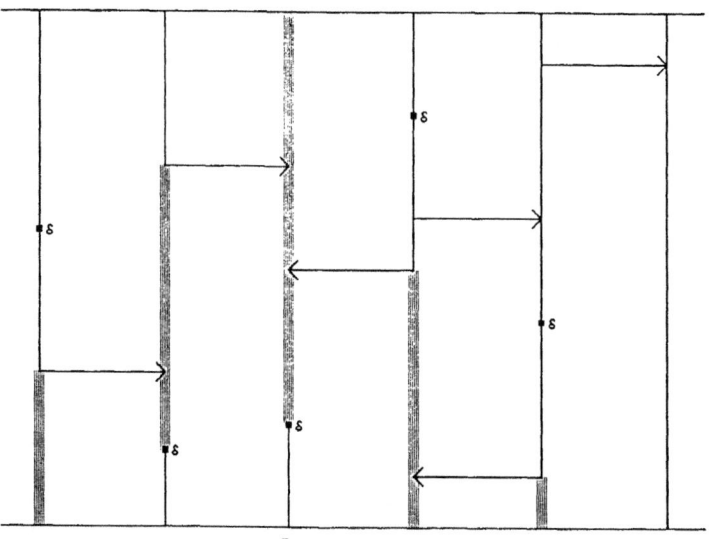

Figure 3.2. Dual of the contact process

is small.

Although left continuous versions of Markov processes are not the traditional ones, we will tolerate them in this section since our main goal is to define the dual process and derive the duality relation (3.2), which is a statement about the one dimensional distributions. (Note that there are only countably many jumps so the left and right continuous versions are equal almost surely at any fixed t.) To construct the dual process starting from time t, we say that there is a *path down from* (y, t) *to* $(x, t - r)$ if there is a sequence of times $s_0 = 0 < s_1 < s_2 < s_n < s_{n+1} = r$ and spatial locations $x_0 = y, x_1, \ldots, x_n = x$ so that

(i) for $i = 1, 2, \ldots, n$ there is an arrow from x_i to x_{i-1} at time $t - s_i$

(ii) the vertical segments $\{x_i\} \times (t - s_{i+1}, t - s_i)$, $i = 0, 1, \ldots n$ do not contain any δ's.

That is, we have to avoid δ's as before but this time we move across arrows in a direction opposite to their orientation. (Exercise: Find a path down from $(3, t)$ to $(2, 0)$ in Figure 3.1.)

The last definition is chosen so that there is a path from $(x, 0)$ to (y, t) if and only if there is a path down from (y, t) to $(x, 0)$ and hence if we define

$$(3.2) \qquad \hat{\xi}_s^{(B,t)} = \{x : \text{ for some } y \in B \text{ there is a path down from } (y, t) \text{ to } (x, t - s)\}$$

then $\{\xi_t^A \cap B \neq \emptyset\} = \{A \cap \hat{\xi}_t^{(B,t)} \neq \emptyset\}$. With a little more thought one sees that for any $0 \leq s \leq t$

$$(3.3) \qquad \{\xi_t^A \cap B \neq \emptyset\} = \{\xi_s^A \cap \hat{\xi}_{t-s}^{(B,t)} \neq \emptyset\} = \{A \cap \hat{\xi}_t^{(B,t)} \neq \emptyset\}$$

Figure 3.2 shows a picture of the dual process $\hat{\xi}_s^{(\{0\},t)}$. To work with the dual, it is useful to define a process $\hat{\xi}_t^B$ so that for each t, $\{\hat{\xi}_s^B; 0 \leq s \leq t\}$ has the same distribution as $\{\hat{\xi}_s^{(B,t)} : 0 \leq s \leq t\}$. Comparing the definition of the original process and the dual shows that we can do this by reversing the direction of the arrows in the original percolation substructure and then applying the original definition. From this observation it should be clear that if ξ_t^A is a contact process with neighborhood set $\mathcal{N}$ then $\hat{\xi}_t^B$ is a contact process with neighborhood set $-\mathcal{N} = \{-x : x \in \mathcal{N}\}$. So if we use our favorite neighborhood $\mathcal{N} = \{x : \|x\|_p \leq r\}$ then the contact process is *self-dual*, i.e., $\{\hat{\xi}_t^B, t \geq 0\}$ and $\{\xi_t^B, t \geq 0\}$ have the same distribution.

Example 3.2. The voter model. Recall that our simple minded voters have two opinions 0 or 1, and that a voter at x changes her opinion at a rate equal to the number of neighbors (i.e., y with $y - x \in \mathcal{N}$) with the opposite opinion. To make the percolation substructure we let $\{U_n^{x,y} : n \geq 1\}$ be independent Poisson processes with rate 1 when $x, y \in \mathbf{Z}^d$ with $y - x \in \mathcal{N}$, we draw an arrow from $(y, U_n^{x,y})$ to $(x, U_n^{x,y})$ and write a δ at $(x, U_n^{x,y})$. We define paths as before and use the paths to define a set valued process in which the state at time t is the set of sites with opinion 1. Writing 1 for occupied and 0 for vacant and thinking about the defintion it is easy to see that the effect of an "arrow-delta" from y to x is as follows:

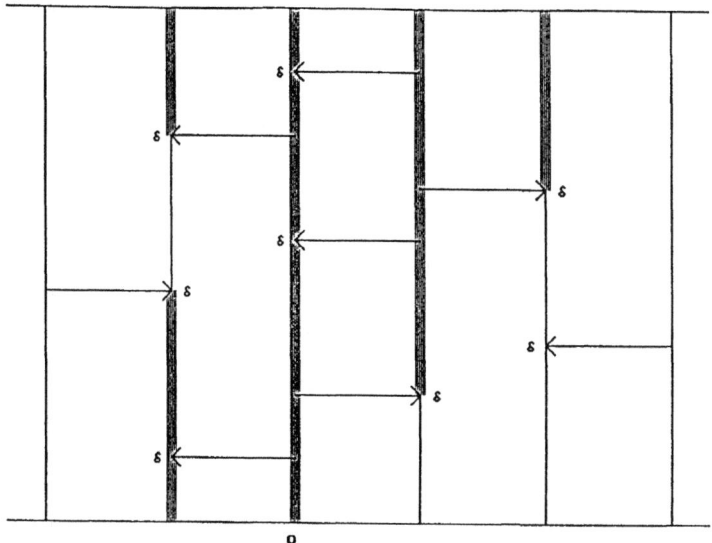

Figure 3.3. Voter model

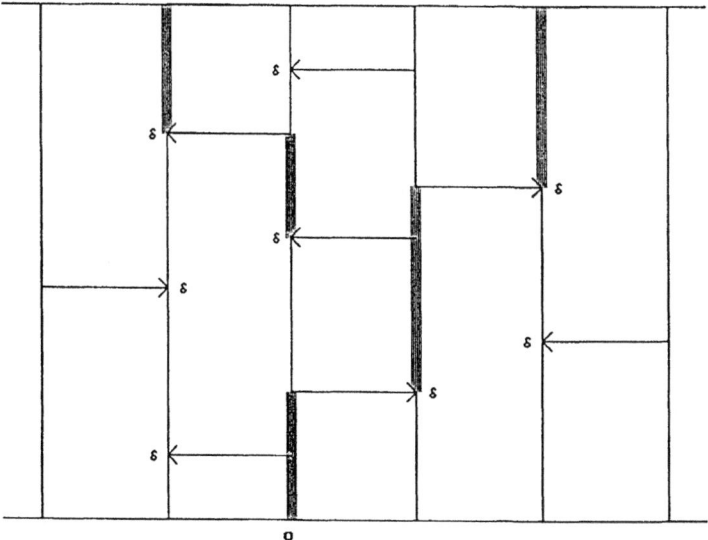

Figure 3.4. Dual of the voter model

	before		after	
	x	y	x	y
$\delta \longleftarrow$	0	0	0	0
$\times \qquad y$	1	0	0	0
	0	1	1	1
	1	1	1	1

In words, because of the δ at x, x will occupied after the "arrow-delta" if and only if y is occupied. From the table (or from the verbal description) we see that the effect of an "arrow-delta" from y to x is to force the voter at x to imitate the voter at y, so the process defined by (3.1) is the voter model. Figure 3.3 gives an example of the construction with $\xi_0 = \{-1, 0\}$. Again the thick lines indicate occupied sites.

The motivation for this construction is that it allows us to define a dual process which in the case of the voter model is quite simple. Since dual paths cannot continue through δ's and can only move across arrows in a direction opposite their orientation, it is easy to check that $\hat{\xi}_s^{(\{x\},t)}$ is always a single site $S_s^{x,t}$, which has the interpretation that the voter at x at time t has the same opinion of the voter at $S_s^{x,t}$ at time $t - s$. See Figure 3.4 which shows $\hat{\xi}_s^{(\{x\},t)}$ for $x = -1$ and $x = 2$. In words, $S_s^{x,t}$ sits at a site y until $t - s = U_n^{y,z}$ for some z, indicating the voter at y imitated the one at z, at which time $S_s^{x,t}$ jumps from y to z. From the last description it should be clear that $S_s^{x,t}$ is a continuous time random walk that for each $w \in \mathcal{N}$ jumps from y to $y + w$ at rate 1.

To determine the behavior of the dual starting from more than one point, we note that it is constructed from a percolation structure with independent Poisson processes $\{U_n^{x,y} : n \geq 1\}$ for $x, y \in \mathbf{Z}^d$ with $y - x \in \mathcal{N}$ at which time we draw an arrow from $(x, U_n^{x,y})$ to $(y, U_n^{x,y})$ and write a δ at $(x, U_n^{x,y})$. From the definition it is easy to see that a "delta-arrows" from x to y has the following effect

	before		after	
	x	y	x	y
$\delta \longrightarrow$	0	0	0	0
$\times \qquad y$	1	0	0	1
	0	1	0	1
	1	1	0	1

The δ at x makes it vacant while the arrow from x to y will make y occupied if there was a particle at y or at x. These are the transitions of a *coalescing random walk*. Particles move independently until they hit and then move together after that. The duality relationship (3.3) between the voter model and coalescing random walks leads easily to the results of Holley and Liggett (1975). These conclusions are true quite generally but we will state them only for our favorite neighborhoods $\{z : \|z\|_p \leq r\}$ with $r \geq 1$. To make the statements here match Theorems 2A and 2B in Section 1, we revert to coordinate notation: $\xi_t(x) = 1$ if and only if $x \in \xi_t$.

Theorem 3.1. *Clustering* occurs in $d \leq 2$. That is, for any ξ_0 and $x, y \in \mathbf{Z}^d$ we have

$$P(\xi_t(x) \neq \xi_t(y)) \to 0 \text{ as } t \to \infty$$

Theorem 3.2. Let ξ_t^θ denote the process starting from an initial state in which the events $\{\xi_0^\theta(x) = 1\}$ are independent and have probability θ. In $d \geq 3$ as $t \to \infty$, $\xi_t^\theta \Rightarrow \xi_\infty^\theta$, a translation invariant stationary distribution in which $P(\xi_\infty(x) = 1) = \theta$.

PROOF OF THEOREM 3.1. From our discussion of the dual it should be clear that

$$P(\xi_t(x) \neq \xi_t(y)) \leq P(S_t^{(x,t)} \neq S_t^{(y,t)})$$

since if the two sites x and y trace their opinions back to the same site at time 0 then they will certainly be equal at time t. Now the difference $S_s^{(x,t)} - S_s^{(y,t)}$ is a random walk stopped when it hits 0, and the random walk has jumps that have mean 0 and finite variance. Such random walks are *recurrent*, and since ours is also an irreducible Markov chain, it will eventually hit 0. Since 0 is an absorbing state for $S_s^{(x,t)} - S_s^{(y,t)}$ it follows that $P(S_t^{(x,t)} \neq S_t^{(y,t)}) \to 0$ and the proof is complete. □

Remark. The reader should not misinterpret Theorem 3.1 as saying that the voter model is boring in $d \leq 2$. Cox and Griffeath (1986) have proved a number of interesting results about the clustering in $d = 2$, which is rather exotic since two dimensional random walk is just barely recurrent.

PROOF OF THEOREM 3.2. From the proof of (2.8) we see that it is enough to prove the convergence of $P(\xi_t \cap B = \emptyset)$ for each B. To treat these probabilities we observe that

$$P(\xi_t \cap B = \emptyset) = E\{(1-\theta)^{|\hat{\xi}_t^{(B,t)}|}\}$$

since by duality there are no particles in B at time t if and only if none of the sites in $\hat{\xi}_t^{(B,t)}$ is occupied at time 0, an event with probability $(1-\theta)^{|\hat{\xi}_t^{(B,t)}|}$. To analyze the right hand side we note that $\hat{\xi}_t^{(B,t)}$ has the same distribution as $\hat{\xi}_t^B$ constructed from the percolation substructure that has the directions of all the arrows reversed. Since $\hat{\xi}_t^B$ is a coalescing random walk, $|\hat{\xi}_t^B|$ is a decreasing function of t and has a limit. Since $0 \leq (1-\theta)^{|\hat{\xi}_t^B|} \leq 1$ it follows from the bounded convergence theorem that

$$\lim_{t \to \infty} E\{(1-\theta)^{|\hat{\xi}_t^{(B,t)}|}\} \text{ exists}$$

and the proof is complete. □

Since the ξ_t^θ are translation invariant (by (2.2)), it follows that the limits ξ_∞^θ are.

$$P(x \in \xi_t^\theta) = P(S_t^{x,t} \in \xi_0^\theta) = \theta$$

for all t so $P(x \in \xi_\infty^\theta) = \theta$. Holley and Liggett (1975) showed that the ξ_∞^θ are spatially ergodic and give all the stationary distributions for the voter model. That is, all stationary distributions are a convex combination of the (distributions of the) ξ_∞^θ. For proofs of this result see the original paper by Holley and Liggett (1975) or Chapter V of Liggett (1985).

Using duality we can prove a convergence theorem due to Harris (1976) for a general class of processes that contains the contact process as a special case. We begin by introducing the models we will consider.

Additive processes. For each finite $A \subset \mathbf{Z}^d$ and $x \in \mathbf{Z}^d$ we introduce independent Poisson processes $\{T_n^{x,A}, n \geq 1\}$ and $\{U_n^{x,A}, n \geq 1\}$ with rates $\lambda(A)$ and $\delta(A)$. (To have a finite range interaction, we only allow finitely many of the rates to be nonzero.) At times $T_n^{x,A}$ we draw arrows from $x + z$ to x for all $z \in A$ and there will be a birth if some site in $x + A$ is occupied. At times $U_n^{x,A}$ we write a δ at x, draw arrows from $x + z$ to x for all $z \in A$, and there will be a death at x unless some point in $x + A$ is occupied. The process is then obtained from the percolation substructure by using (3.1). In the new notation our two examples may be written as (the rates we do not mention are 0):

The contact process. $\lambda(A) = \lambda$ if $A = \{x\}$ with $x \in \mathcal{N}$; $\delta(\emptyset) = 1$.

The voter model. $\delta(A) = 1$ if $A = \{x\}$ with $x \in \mathcal{N}$.

It should be clear that for any additive process the birth rates are increasing and the death rates are decreasing so these systems are attractive. To see that additive processes are a fairly small subclass of the attractive models, we will now consider

Example 3.3. Nonlinear Contact Processes. In these systems the flip rates are

$$c_0(x, \xi) = 1$$
$$c_1(x, \xi) = b(|\{y \in \mathcal{N} : \xi(x + y) = 1\}|)$$

where $b(0) = 0$. To get the desired death rates we set $\delta(\emptyset) = 1$ and $\delta(A) = 0$ otherwise. To see what birth rates we can create we begin with the special case

(i) $d = 1, \mathcal{N} = \{-1, 1\}$. In this situation we must have

$$\lambda(\{1\}) = \lambda(\{-1\}) = a_1 \qquad \lambda(\{1, -1\}) = a_2$$

and the other $\lambda(A) = 0$, so $b(1) = a_1 + a_2$ and $b(2) = 2a_1 + a_2$ which is possible with $a_1, a_2 \geq 0$ if and only if

$$b(1) \leq b(2) \leq 2b(1)$$

The extreme case $b(2) = 2b(1)$ is the basic contact process, the other extreme $b(2) = b(1) = b$ is called the *threshold contact process* because the birth rate is b if there is at least one occupied neighbor. An example of a system not covered by this construction is the *sexual reproduction model* which has $b(1) = 0$ and $b(2) = \lambda$.

(ii) Suppose $|\mathcal{N}| = 4$ and think about $\mathcal{N} = \{-2, -1, 1, 2\}$ in $d = 1$ or $\mathcal{N} = \{z : \|z\|_1 = 1\}$ in $d = 2$. (The geometry of the set $\mathcal{N}$ does not enter into the decision as to whether or not a system is additive.) In this case $\lambda(A) = a_i$ if $A \subset \mathcal{N}$ with $|A| = i$ (and 0 otherwise) so

$$b(1) = a_1 + 3a_2 + 3a_3 + a_4$$
$$b(2) = 2a_1 + 5a_2 + 4a_3 + a_4$$
$$b(3) = 3a_1 + 6a_2 + 4a_3 + a_4$$
$$b(4) = 4a_1 + 6a_2 + 4a_3 + a_4$$

To see the equation of $b(2)$ say, note that any two element subset of $\mathcal{N}$ touches 2 of the singleton subsets of $\mathcal{N}$, all but one of the 6 two element subsets, all 4 of the three element subsets, and the four element subset. Subtracting the equations gives

$$b(4) - b(3) = a_1$$
$$b(3) - b(2) = a_1 + a_2$$
$$b(2) - b(1) = a_1 + 2a_2 + a_3$$
$$b(1) - b(0) = a_1 + 3a_2 + 3a_3 + a_4$$

and taking differences again

$$a_1 = b(4) - b(3)$$
$$a_2 = (b(3) - b(2)) - (b(4) - b(3))$$
$$a_3 = (b(2) - b(1)) - 2(b(3) - b(2)) + (b(4) - b(3))$$
$$a_4 = ((b(1) - b(0)) - 3(b(2) - b(1)) + 3(b(3) - b(2)) - (b(4) - b(3))$$

The process is additive if and only if these quantities are nonnegative. These conditions are monotonicity and convexity properties of the sequence of birth rates $b(i)$. A result for general neighborhoods can be found in Harris (1976), see (6.4) on page 184. The conclusions we would like the reader to draw from this computation are that (i) the additive processes are a small subset of the attractive processes but (ii) when we consider nonlinear contact processes with $|\mathcal{N}| = 4$ additive processes are a four dimensional subset of the four dimensional set of models.

Harris' convergence theorem for additive processes. Before getting started we need to introduce a technical condition. Let ξ_t^0 denote the process starting from a single particle at the origin. We say ξ_t is *irreducible* if for any x and $t > 0$ $P(x \in \xi_t) > 0$. Recall that in Section 2, we let ξ_t^1 denote the process starting from $\xi_0^1 = \mathbf{Z}^d$ and showed that for any attractive process $\xi_t^1 \Rightarrow \xi_\infty^1$, a translation invariant stationary distribution.

Theorem 3.3. Suppose ξ_t is an irreducible additive process with $\delta(\emptyset) > 0$. If ξ_0 is translation invariant and assigns 0 probability to the empty configuration then $\xi_t \Rightarrow \xi_\infty^1$ as $t \to \infty$.

Corollary. ξ_∞^1 is the only translation invariant stationary distribution that assigns 0 probability to the empty configuration.

Remarks. The condition $\delta(\emptyset) = 0$ eliminates the voter model for which the conclusion of Theorem 3.3 is always false. Our result is only for translation invariant initial distributions. With a lot more work one can prove a *complete convergence theorem*:

Theorem 3.4 Suppose ξ_t is an irreducible additive process with $\delta(\emptyset) > 0$. Then for any A,

$$\xi_t^A \Rightarrow P(\tau^A < \infty)\delta_\emptyset + P(\tau^A = \infty)\xi_\infty^1$$

where $\delta_\emptyset$ denotes the pointmass on the emptyset and we are using ξ_∞^1 to denote its distribution.

In words, if the process does not die out, then at large times it looks like the process starting from all 1's. This implies that all stationary distributions have the form $\theta\delta_\emptyset + (1 - \theta)\xi_\infty^1$. For the contact process, this result is due to Bezuidenhout and Grimmett (1990). To prove this in the general case you will need to consult Bezuidenhout and Gray (1993).

PROOF OF THEOREM 3.3. To begin we note that the duality equation (3.3) implies

$$P(\xi_t^1 \cap B \neq \emptyset) = P(\hat{\xi}_t^{(B,t)} \cap \xi_0^1 \neq \emptyset)$$
$$= P(\hat{\xi}_t^B \neq \emptyset) \to P(\hat{\tau}^B = \infty)$$

as $t \to \infty$. As in the proof of Theorem 3.2, the argument in (2.8) shows that it is enough to prove $P(\xi_t \cap B \neq \emptyset) \to P(\hat{\tau}^B = \infty)$. Half of this is very easy. By duality and the fact that $\xi_0 \subset \mathbf{Z}^d$

$$P(\xi_t \cap B \neq \emptyset) = P(\xi_0 \cap \hat{\xi}_t^{(B,t)} \neq \emptyset) \leq P(\hat{\tau}^B > t)$$

so

$$\limsup_{t\to\infty} P(\xi_t \cap B \neq \emptyset) \leq P(\hat{\tau}^B = \infty)$$

To prove the other direction, we let t_0 be the constant in (2.1) and observe that (3.3) implies

$$P(\xi_{t+t_0} \cap B \neq \emptyset) = P(\xi_{t_0} \cap \hat{\xi}_t^{(B,t+t_0)} \neq \emptyset)$$

To get the right hand side to converge to $P(\hat{\tau}^B = \infty)$ we need to show that when $\hat{\xi}_t^{(B,t+t_0)} \neq \emptyset$ then it will intersect ξ_{t_0} with high probability. The first step in doing this is to show that when $\hat{\xi}_t^{(B,t+t_0)} \neq \emptyset$, it will contain a large number of points with high probability. To do this, let

$$\Lambda = \sum_A |A|(\lambda(A) + \delta(A))$$

be the rate at which an isolated particle gives birth to a new particle and let $\alpha = (1 - e^{-\delta(\emptyset)})e^{-\Lambda}$ be a lower bound on the probability that in one unit of time an isolated particle is killed and does not give birth. Now for any K

$$P(t < \hat{\tau}^B \leq t+1) \geq \alpha^K P(0 < |\hat{\xi}_t^{(B,t+t_0)}| \leq K)$$

To see this note that the events that each particle is killed by a δ are independent, and write the statement that no particle gives birth in terms of Poisson processes in the percolation substructure. Since $P(t < \hat{\tau}^B \leq t+1) \to 0$ as $t \to \infty$, and α^K is a positive constant, it follows that

(3.4) $$P(0 < |\hat{\xi}_t^{(B,t+t_0)}| \leq K) \to 0$$

To complete the proof now it suffices to show

(3.5) Lemma. If $\epsilon > 0$ then we can pick K large enough so that if $|A| \geq K$ then $P(\xi_{t_0} \cap A = \emptyset) \leq 3\epsilon$.

For then it follows that from (3.5) and (3.4) that

$$\liminf_{t\to\infty} P(\xi_{t_0} \cap \hat{\xi}_t^{(B,t+t_0)} \neq \emptyset) \geq (1 - 3\epsilon)\liminf_{t\to\infty} P(|\hat{\xi}_t^{(B,t+t_0)}| \geq K)$$

$$\geq (1 - 3\epsilon)P(\hat{\tau}^B > t)$$

Remark. For the conclusion in (3.5) it is important that we let the process run for a positive amount of time. The initial configuration ξ_0 that is $2\mathbf{Z}$ with probability $1/2$ and $2\mathbf{Z} + 1$ with probability $1/2$ is translation invariant but $P(\xi_0 \cap \{2,4,\ldots,2K\}) = 1/2$ for all K.

PROOF OF (3.5): For this proof it is convenient to use the coordinate representation of the process, i.e., $\xi_t(x) = 1$ if x is occupied at time t and 0 otherwise. Let μ be the distribution of ξ_0 (i.e., the induced measure on $\{0,1\}^S$) and use P_ξ to denote the probability law for ξ_t when $\xi_0 = \xi$. Our assumption of irreducibility and attractiveness imply that $P_\xi(\xi_{t_0}(x) = 1) > 0$ unless $\xi \equiv 0$, an event that by assumption has probability 0, so

(3.6) For any $\epsilon > 0$ there is a $\rho < 1$ so that

$$\mu(\{\xi : P_\xi(\xi_{t_0}(x) = 0) > \rho\}) \leq \epsilon$$

Here we need translation invariance to conclude that the left hand side does not depend on x. The second ingredient is to note repeated use of Hölder's inequality gives

$$E(X_1 \cdots X_k) \leq (E|X_1^k|)^{1/k} \cdots (E|X_k^k|)^{1/k}$$

which in turn implies

(3.7) Let $X_1,\ldots,X_k$ be random variables so that $0 \leq X_i \leq 1$ and $P(X_i > \rho) \leq \epsilon$. Then

$$E(X_1 \cdots X_k) \leq \rho^k + \epsilon$$

Pick J so that $\rho_\epsilon^J \leq \epsilon$. Our proof of the next result explains why we chose the time t_0. The result is valid for any time t, see Holley (1972).

(3.8) Given $\epsilon > 0$ and J, we can pick L so that if $B \subset \mathbf{Z}^d$ with $|B| = J$ and $\|x - y\|_\infty > 2L$ whenever $x, y \in B$ with $x \neq y$ then

$$\left| E_\xi \left\{ \prod_{x \in B} (1 - \xi_{t_0}(x)) \right\} - \prod_{x \in B} \{E_\xi(1 - \xi_{t_0}(x))\} \right| \leq \epsilon$$

PROOF OF (3.8): First we compute the value of each $\xi_{t_0}(x)$ with $x \in B$ by using an independent copy of the percolation substructure $\mathcal{P}_x$. The second step is to combine all these independent substructures to make a new one $\mathcal{P}_{all}$ by taking $T_n^{y,A}$ and $U_n^{y,A}$ from $\mathcal{P}_x$ if and only if $y + A \subset D(x, L) = \{z : \|x - z\|_\infty \leq L\}$ and then using another independent percolation substructure $\mathcal{P}^*$ to fill in the missing Poisson processes. Let R be the largest value of $\|x\|_\infty$ for a point in some set A with $\lambda(A)$ or $\delta(A) > 0$. R is the range of the interaction. If the cluster containing x in $\mathcal{P}_x$ defined in the proof of (2.1) lies inside $D(x, L - R)$ then it is identical with the cluster containing x in $\mathcal{P}_{all}$ and the values computed for ξ_{t_0} are the same. Since the states of x in the processes on $\mathcal{P}_x$ are independent, it follows from the proof of (2.1) that if L is large the random variables $1 - \xi_{t_0}(x)$ on $\mathcal{P}_{all}$ are equal with high probability to independent random variables and (3.8) follows. $\quad\square$

To complete the proof of (3.5) now, we observe that

(3.9) If $B \subset \mathbf{Z}^d$ with $|B| = J$ and if $\|x - y\|_\infty > 2L$ whenever $x, y \in B$ with $x \neq y$ then

$$P(\xi_{t_0}(x) = 0 \text{ for all } x \in B) = \int \mu(d\xi) E_\xi \prod_{x \in B}(1 - \xi_{t_0}(x))$$

$$\leq \epsilon + \int \mu(d\xi) \prod_{x \in B} E_\xi(1 - \xi_{t_0}(x)) \leq 2\epsilon + \rho_\epsilon^J \leq 3\epsilon$$

by (3.8), (3.6), (3.7), and the choice of J. To get from the last result to the desired conclusion we let $K = (4L+1)^d J$ and observe that if $|A| \geq K$ we can find a subset B with $|B| = J$ that sastisfies the hypotheses of (3.9). $\quad\square$

Example 3.4. Multitype contact processes, defined in Section 1, have state space $\{0, 1, \kappa - 1\}^S$ where 0 indicates a vacant site and $i > 0$ indicates a site occupied by one plant of type i, and have flip rates that are linear:

$$c_0(x, \xi) = \delta_{\xi(x)}$$
$$c_i(x, \xi) = \lambda_i n_i(x, \xi) \qquad \text{if } \xi(x) = 0$$

When $\lambda_i = \lambda$ and $\delta_i = \delta$, this process can be studied by using a duality that is a hybrid of the one for the contact process and for the voter model. The first step is to construct the process as we did the contact process. We introduce independent Poisson processes $\{U_n^x, n \geq 1\}$ with rate δ and $\{T_n^{x,y}, n \geq 1\}$ with rate λ for each $x, y \in \mathbf{Z}^d$ with $y - x \in \mathcal{N}$. As before, we write a δ at (x, U_n^x) to indicate that a death will occur if x is occupied by a particle of either type, and we draw an arrow from $(y, T_n^{x,y})$ to $(x, T_n^{x,y})$ to indicate that if x is vacant and y is occupied then there will be a birth from y to x.

If we define the dual process as in (3.2) then reasoning as before we see that x will be occupied at time t if and only if some site in $\hat{\xi}^{(\{x\},t)}_t$ is occupied in ξ_0. The dual for the mutlitype contact process is the set $\hat{\xi}^{(\{x\},t)}_t$ plus an ordering of that set with the interpretation that the type of x is that of the first occupied site in the ordering. For example in the realization drawn in Figure 3.2, the ordering is $1 > 2 > -2$

The first site in $\hat{\xi}^{(\{x\},t)}_s$ in this ordering is called the *distinguished particle*. Results of Neuhauser (1992) show that the movements of the distinguished particle are enough like those of a random walk to conclude that in $d \leq 2$ the distinguished particles for the duals of two different sites will eventually be equal for large t. This is the key idea in proving Theorems 4C and 4D in Section 1. In the two type case, when $\delta_1 = \delta_2$ and $\lambda_1 < \lambda_2$ we can augment the construction above with Poisson processes of arrows that only allow the births of 2's and an easy argument gives Theorem 4A. However such an approach will never give us Conjecture 4B.

4. A Comparison Theorem

In this section we will introduce a comparison theorem that is very useful in proving the existence of nontrivial translation invariant stationary distributions. At this point we have to ask for the reader's patience: the result given in Theorem 4.3 is powerful but you will need to see a few applications to understand how it works.

Our general method for proving the existence of stationary distributions is to compare the process of interest with oriented percolation, so our first step is to introduce oriented percolation and state some of its basic properties, the proofs of which are hidden away in the appendix. Let

$$\mathcal{L}_0 = \{(x, n) \in \mathbf{Z}^2 : x + n \text{ is even}, n \geq 0\}$$

and make $\mathcal{L}_0$ into a graph by drawing oriented edges from (x, n) to $(x + 1, n + 1)$ and from (x, n) to $(x - 1, n + 1)$. Given random variables $\omega(x, n)$ that indicate whether the sites are open (1) or closed (0), we say that (y, n) can be reached from (x, m) and write $(x, m) \to (y, n)$ if there is a sequence of points $x = x_m, \ldots, x_n = y$ so that $|x_k - x_{k-1}| = 1$ for $m < k \leq n$ and $\omega(x_k, k) = 1$ for $m \leq k \leq n$. In the standard oriented percolation model the variables $\omega(x, n)$ are independent, but in almost all cases our comparisons will introduce dependencies between the $\omega(x, n)$, so we need a more general set-up. We say that the $\omega(x, n)$ are "M *dependent with density at least* $1 - \gamma$" if whenever (x_i, n_i), $1 \leq i \leq I$ is a sequence with $\|(x_i, n_i) - (x_j, n_j)\|_\infty > M$ if $i \neq j$ then

$$(4.1) \qquad P(\omega(x_i, n_i) = 0 \text{ for } 1 \leq i \leq I) \leq \gamma^I$$

Note: Classical M-dependence would require that the $\omega(x_i, n_i)$ considered above are independent. However the probability in (4.1) is the only one we need to control and hence the only thing we assume.

Given an initial condition $W_0 \subset 2\mathbf{Z} = \{x : (x, 0) \in \mathcal{L}_0\}$, we can define a process by

$$W_n = \{y : (x, 0) \to (y, n) \text{ for some } x \in W_0\}$$

In words, the sites W_n are those that are wet at time n. To keep the terminology straight, think of open sites as air spaces in a rock, and the sites in W_n as the ones that the fluid can reach (and hence wet) at level n. We use W_n^0 to denote the process that results when $W_0^0 = \{0\}$ and we let

$$\mathcal{C}_0 = \{(y, n) : (0, 0) \to (y, n)\}$$

be the set of all points in space-time that can be reached by a path from $(0, 0)$. (When $(0, 0)$ is open $\mathcal{C}_0 = \cup_n(W_n^0 \times \{n\})$.) $\mathcal{C}_0$ is called the *cluster containing the origin*. Figure 4.1 shows a simulation of the independent oriented percolation process in which sites are open (indicated by black dots) with probability $p = 0.6$. Time goes up the page and lines connect the points of $\mathcal{C}_I$.

When the cluster containing the origin is infinite, i.e., $\{|\mathcal{C}_0| = \infty\}$ we say that *percolation occurs*. Our first result shows that if the density of open sites is high enough then percolation occurs. All that is important about the upper bound is that it is < 1 for small γ and converges to 0 as $\gamma \to 0$.

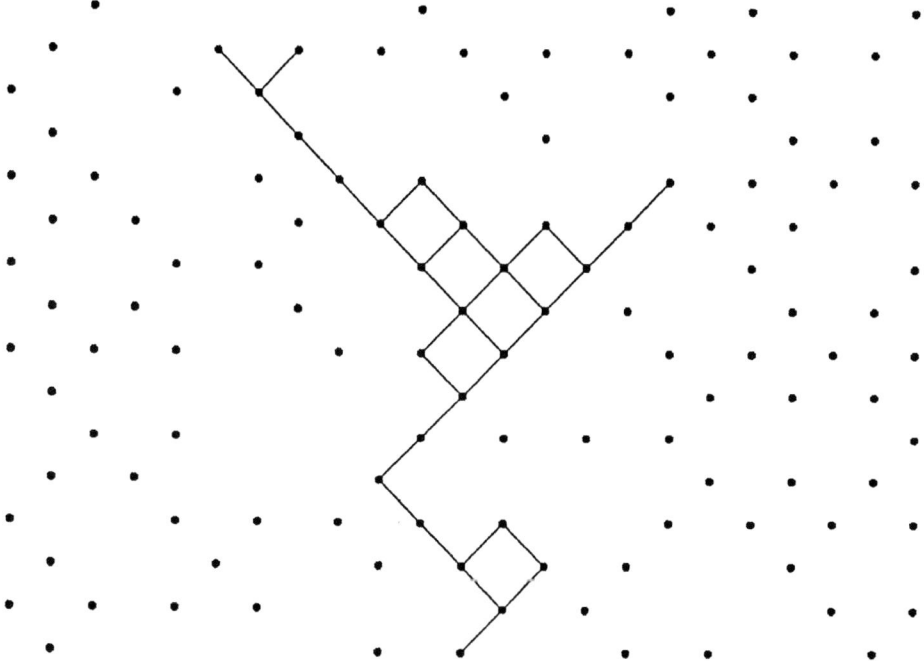

Figure 4.1

Theorem 4.1. If $\gamma \leq 6^{-4(2M+1)^2}$ then

$$P(|\mathcal{C}_0| < \infty) \leq 55\,\theta^{1/(2M+1)^2} \leq 1/20$$

In order to prove the existence of stationary distributions we need results about M dependent oriented percolation starting from the initial configuration W_0^p in which the events $\{x \in W_0^p\}$, $x \in 2\mathbf{Z}$ are independent and have probability p. We will sometimes call this a *Bernoulli random set with density p*. Taking $p = 1$ (i.e., all sites wet initially) corresponds to computing the upper invariant measure for oriented percolation, but for some of the proofs below we will need to allow $p < 1$. Note that the estimate on the lim inf is independent of p and is 1 minus the upper bound in Theorem 4.1.

Theorem 4.2. If $p > 0$ and $\gamma \leq 6^{-4(2M+1)^2}$ then

$$\liminf_{n \to \infty} P(0 \in W_{2n}^p) \geq 1 - 55\,\gamma^{1/(2M+1)^2} \geq 19/20$$

The last result shows that if the density of open sites in oriented percolation is sufficiently high and if we start with from a Bernoulli random set with density p then the probability 0 is wet at time t does not go to 0. This result will allow us to prove in a number of situations that if we start from a suitably chosen translation invariant initial distribution, then the density of sites of type i does not go to 0 and then using (2.7) or (2.11) that a nontrivial translation invariant stationary distribution exists. The missing link is provided by Theorem 4.3, which gives general conditions that guarantee a process dominates oriented percolation. This is the result we warned the reader about at the beginning of the section – it does not look pretty but it is very useful in a number of situations.

Comparison Assumptions. We suppose given the following ingredients: a translation invariant finite range process $\xi_t : \mathbf{Z}^d \to \{0, 1, \ldots \kappa - 1\}$ that is constructed from the graphical representation given in Section 2, an integer L, and a collection H of configurations determined by the values of ξ on $[-L, L]^d$ with the following property:

if $\xi \in H$ then there is an event G_ξ measurable with respect to the graphical representation in $[-k_0 L, k_0 L]^d \times [0, j_0 T]$ and with $P(G_\xi) \geq (1 - \gamma)$ so that if $\xi_0 = \xi$ then on G_ξ, ξ_T lies in $\sigma_{2Le_1} H$ and in $\sigma_{-2Le_1} H$.

Here $(\sigma_y \xi)(x) = \xi(x + y)$ denotes the translation (or shift) of ξ by y and $\sigma_y H = \{\sigma_y \xi : \xi \in H\}$. If we let $M = \max\{j_0, k_0\}$ then the space time regions

$$\mathcal{R}_{m,n} = (m 2Le_1, nT) + \left\{ [-k_0 L, k_0 L]^d \times [0, j_0 T] \right\}$$

that correspond to points $(m, n), (m', n') \in \mathcal{L}$ with $\|(m, n) - (m', n')\|_\infty > M$ are disjoint.

For a concrete instance of the comparison assumptions consider the applications we will make to the threshold contact process in Section 5 and to the basic contact process in Section 7. In both cases $\kappa = 2$, and H is the set of configurations with at least K 1's in $[-L, L]^d$, $k_0 = 4$, and $j_0 = 1$. In words, we show that if there is a "pile" of at least K particles in $[-L, L]^d$ then with high probability there will be piles of at least K particles in $-2Le_1 + [-L, L]^d$ and in $2Le_1 + [-L, L]^d$ at time T, and the event that guarantees this is measurable with respect to the graphical representation in $[-4L, 4L]^d \times [0, T]$. Figure 4.2 below gives a picture of the event.

Using words inspired by the contact process example, our comparison assumptions say that if we have a "pile of particles" in $I_m = m 2Le_1 + [-L, L]^d$ at time nT (i.e., $\xi_{nT} \in \sigma_{m 2Le_1} H$) then with high probability we will have piles of particles in I_{m-1} and in I_{m-1} at time $(n + 1)T$, and the event that guarantees this is measurable with respect to the graphical representation in $\mathcal{R}_{m,n}$. If we think of drawing arrows from (m, n) to $(m + 1, n + 1)$ and to $(m - 1, n + 1)$ whenever the good event in $\mathcal{R}_{m,n}$ occurs then the connection with oriented percolation should be clear.

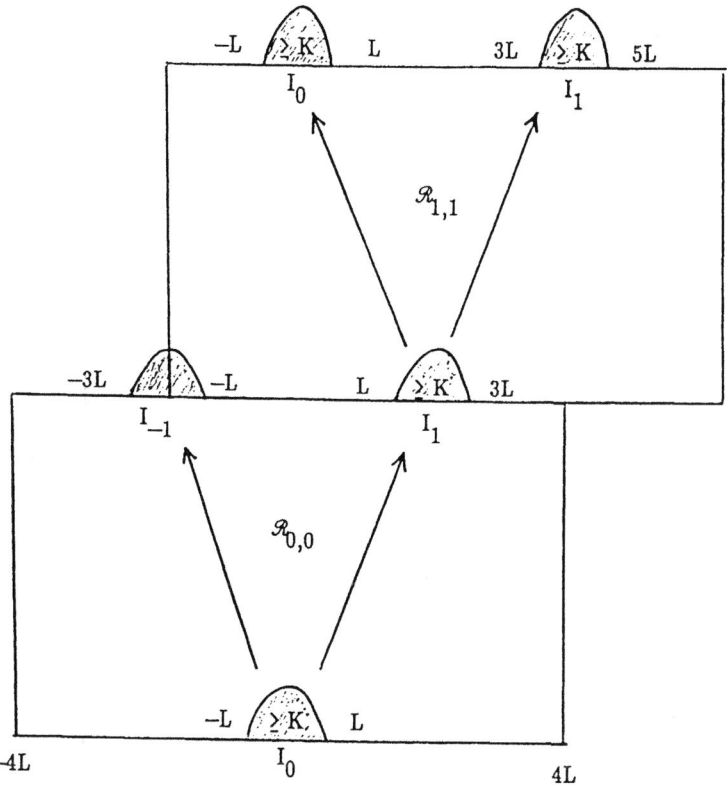

Figure 4.2

To formulate our theorem we let $X_n = \{m : (m, n) \in \mathcal{L}_0, \xi_{nT} \in \sigma_{m2Le_1} H\}$. Intuitively, $m \in X_n$ if there is a pile of particles in I_m at time nT

Theorem 4.3. If the comparison assumptions hold then we can define random variables $\omega(x, n)$ so that X_n dominates an M dependent oriented percolation process with initial configuration $W_0 = X_0$ and density at least $1 - \gamma$, i.e., $X_n \supset W_n$ for all n.

Again the details are hidden away in the appendix so that they can be digested after the reader has seen that this is a useful result.

Our first indication that Theorems 4.1–4.3 are useful is a simple proof of a general result about the existence of stationary distributions, which contains as a special case a number of earlier results. To formulate our result we will consider a fixed set of increasing birth rates $c_1(x, \xi)$ and introduce death rates $c_0(x, \xi) \equiv \epsilon$. We say that the birth rates are *robust* if there is an $\epsilon_0 > 0$ so that there is a translation invariant stationary distribution with a positive density of 1's for $\epsilon < \epsilon_0$. Our next result gives a sufficient condition for robustness. It may look a little strange at first but it has been formulated to be easy to prove and to apply.

Theorem 4.4. Let $\bar{\xi}_t^{L,\rho}$ denote the process with no deaths, i.e., $\epsilon = 0$, starting from $\bar{\xi}_0^{L,\rho}(x) = 1$ for $x \in [-L, L]^d$, $= 0$ otherwise and modified so that no births are allowed outside $[-\rho L, \rho L]^d$. Suppose that we can pick $\rho \geq 3$ so that for any $\delta > 0$ we can pick L and $T < \infty$ so that

$$P(\bar{\xi}_T^{L,\rho}(x) = 1 \text{ for all } x \in [-3L, 3L]^d) \geq 1 - \delta$$

Then the birth rates are robust (and fertile).

Ignoring the undefined term in parentheses, this theorem says that if, in the absence of deaths, the birth mechanism can triple the size of a cube $[-L, L]^d$ with high probability, then there is a nontrivial translation invariant stationary distribution when the death rate $c_0(x, \xi) \equiv \epsilon$ is small. The requirement that this can be done when the model is "modified so that no births are allowed outside $[-\rho L, \rho L]^{d}$" is a technical condition that is usually satisfied with $\rho = 3$.

PROOF OF THEOREM 4.4: If we let $K = \rho$, $J = 1$ and $H = \{\xi : \xi(x) = 1 \text{ for all } x \in [-L, L]^d\}$. then the hypotheses of Theorem 4.4 are that the comparison assumptions hold for the system with $\epsilon = 0$. However, once L and T are fixed it follows that for $\epsilon \leq \epsilon_0$, the good event G_ξ for the one configuration in H has probability at least $1 - 2\delta$, since the probability a death occurs at some site in the space time box $[-3L, 3L]^d \times [0, T]$ is less than δ when ϵ_0 is sufficiently small.

To construct our stationary distribution, we consider the process ξ_t^1 starting from $\xi_0^1(x) = 1$ for all x. In this case $X_0 = 2\mathbf{Z}$ so using Theorems 4.3 and 4.2 with $p = 1$, it follows that if $\epsilon \leq \epsilon_0$ then

$$\liminf_{n \to \infty} P(\xi_{nT}^1(0) = 1) \geq 19/20$$

Using (2.7) now it follows that there is a nontrivial stationary distribution. □

We will now give three examples to shows that is easy to check the conditions of Theorem 4.4.

Corollary 4.5. If we fix $\lambda = 1$ in the contact process with neighborhood $\mathcal{N} = \{x : \|x\|_p \leq r\}$ where $r \geq 1$ then there is a nontrivial stationary distribution when the death rate $\delta < \delta_0$.

PROOF: Take $\rho = 3$ and $L = 1$. Since 1's can never flip to 0 it is easy to see that

$$\lim_{T \to \infty} P(\bar{\xi}_T^{L,\rho}(x) = 1 \text{ for all } x \in [-3L, 3L]^d) = 1$$

so the hypotheses of Theorem 4.4 are satisfied. □

Example 4.1. One Dimensional Counting Rules. Suppose $d = 1$, $\mathcal{N} = \{z : |z| \le k\}$, and let

$$n_1(x, \xi) = |\{z \in \mathcal{N} : \xi(x + z) = 1\}|$$

be the number of neighbors of x that are 1. We call a birth rate $c_1(x, \xi)$ a *counting rule* if it only depends on the number of 1's in the neighborhood, i.e., $c_1(x, \xi) = b(n_1(x, \xi))$ Clearly a counting rule birth rate is increasing if and only if $j \to b(j)$ is nondecreasing. Let $j_0 = \min\{j : b_j > 0\}$ and call j_0 the *order* of the birth rate. The next result is due to Mityugin.

Corollary 4.6. When $d = 1$ and $\mathcal{N} = \{j : |j| \le k\}$, increasing counting rule birth rates are robust if and only if their order $j_0 \le k$.

PROOF: If $j_0 > k$ then a string of at least $k + 1$ consecutive 0's can never flip back to 1 even if all the other sites are 1. If $c_0(x, \xi) \equiv \epsilon > 0$ then such a string will eventually be created and grow to cover the whole line, so there cannot be a nontrivial stationary distribution.

If $j_0 \le k$, we take $\rho = 3$ and choose L so that $2L + 1 \ge k$. When $\epsilon = 0$ the 1's never flip back to 0. The 0 at $L + 1$ has k neighbors that are 1 and hence flips to 1 at rate $b(k) \ge b(j_0) > 0$. Once the 0 at $L + 1$ flips to 1, the 0 at $L + 2$ will flip to 1 at rate $b(k)$, so

$$\lim_{T \to \infty} P(\bar{\xi}_T^{L,\rho}(x) = 1 \text{ for all } x \in [-3L, 3L]) = 1$$

and the hypotheses of Theorem 4.4 are satisfied. □

Things get more interesting in two dimensions.

Example 4.2. Two Dimensional Threshold Birth Rates. Suppose $d = 2$ and $\mathcal{N} = \{z : \|z\|_\infty = 1\}$, i.e., in addition to the four nearest neighbors we use the four diagonally adjacent points:

$$\mathcal{N} = \begin{cases} (-1, 1) & (0, 1) & (1, 1) \\ (-1, 0) & & (1, 0) \\ (-1, -1) & (0, -1) & (1, -1) \end{cases}$$

This is sometimes called the *Moore neighborhood* in honor of one of the pioneers in the field of cellular automata. Let $n_1(x, \xi) = |\{j \in \mathcal{N} : \xi(x) = 1\}|$ be the number of neighbors in state 1 and let

$$c_1(x, \xi) = \begin{cases} 1 & \text{if } n_1(x, \xi) \ge \theta \\ 0 & \text{if } n_1(x, \xi) < \theta \end{cases}$$

This is called a threshold θ since the birth rate 1 if there are at least θ 1's in the neighborhood then the birth rate is 1, and otherwise it is 0. From Theorem 4.4 we get easily that

Corollary 4.7. Two dimensional threshold birth rates for the Moore neighborhood in two dimension are robust if $\theta \leq 3$.

PROOF: Take $\rho = 3$, $L = 1$, and draw a picture.

$$
\begin{array}{ccccc}
4 & 3 & 2 & 3 & 4 \\
3 & 1 & 1 & 1 & 3 \\
2 & 1 & 1 & 1 & 2 \\
3 & 1 & 1 & 1 & 3 \\
4 & 3 & 2 & 3 & 4
\end{array}
$$

We start with the 3×3 square of 1's occupied by 1's. If $\theta \leq 3$ then the four sites marked with 2's have birth rate 1 and will eventually become occupied. Once they do, the eight sites marked 3 have three occupied neighbors and will become occupied. Finally the four sites marked 4 will become occupied. At this point we have shown how the process can fill up $[-2, 2]^2$. Repeating the argument, it is easy to see that

$$\lim_{T \to \infty} P(\bar{\xi}_T^{L, \rho}(x) = 1 \text{ for all } x \in [-3, 3]^2) = 1$$

the hypothesis of Theorem 4.4 is satisfied and the result follows. □

In the last argument it was important that we used the Moore neighborhood, instead of the usual nearest neighbors $\{z : |z| = 1\}$. If we use the nearest neighbors then, no matter how big L, is if we start with $[-L, L]^2$ occupied nothing happens since any site outside $[-L, L]^2$ has at most one occupied neighbor.

$$
\begin{array}{ccccc}
0 & 0 & 0 & 0 & 0 \\
0 & 0 & x & 0 & 0 \\
1 & 1 & 1 & 1 & 1
\end{array}
$$

Since births are impossible outside any rectangle containing the 1's in the initial configuration, it is clear that the threshold two birth rate for the nearest neighbors *dies out* whenever the death rate is $c_0(x, \xi) \equiv \epsilon > 0$. That is, if there are only finitely many 1's in ξ_0, then

$$P(\xi_t \not\equiv 0) \to 0 \quad \text{as } t \to \infty$$

Here $\xi_t \equiv 0$ is short for $\xi_t(x) = 0$ for all x. Note that the all 0's state is absorbing so $t \to P(\xi_t \not\equiv 0)$ is decreasing. The opposite of dies out is *survives*. That is, if L is large enough and we start with 1's on $[-L, L]^d$ then

$$\lim_{t \to \infty} P(\xi_t \not\equiv 0) > 0 \quad \text{as } t \to \infty$$

We say that a birth rate is *fertile* if it survives when $c_0(x, \xi) = \epsilon$ and $\epsilon < \epsilon_0$. As the parenthetical phrase in Theorem 4.4 indicates, our sufficient conditions for robustness are also sufficient for fertility.

Having two notions of what it means for birth rates to be large enough, fertility and robustness, it is natural ask what is the relationship between these two notions:

1. Results of Bezuidenhout and Gray imply that increasing birth rates that are fertile are also robust, but the two notions are not equivalent.

2. As we have shown the two dimensional threshold two system using the nearest neighbors is not fertile. However, Bramson and Gray (1991) have shown that it is robust. Intuitively the process cannot grow outside of a rectangle but it is good at filling in holes that develop so it can have a nontrivial stationary distribution when ϵ is small.

In the case of the Moore neighborhood in two dimensions, it is easy to see that the threshold 4 system is not fertile but techniques of Bramson and Gray (1991) can be used to show that it is robust. The threshold 5 system has finite configurations of 0's that cannot be filled in

$$
\begin{array}{cccc}
 & 0 & 0 & \\
0 & 0 & 0 & 0 \\
0 & 0 & 0 & 0 \\
 & 0 & 0 & \\
\end{array}
$$

so an easy argument shows that it is not robust. An interesting open problem is to look at the neighborhoods $\mathcal{N} = \{z : \|z\|_p \le r\}$ (or even just take $p = \infty$) and find the largest thresholds for which the threshold θ birth rule on that neighborhood is robust (resp. fertile).

Further results. There are many other results proving the existence of phase transitions for processes with state space $\{0, 1\}^S$. Gray and Griffeath (1982) proved a "stability theorem for attractive nearest neighbor spin systems on $\mathbf{Z}$" by the contour method, a result which was reproved by the methods of this section by Bramson and Durrett (1988). Gray (1987) proved results for the one dimensional majority vote model. Chen (1992) used ideas from bootstrap percolation to study a model with sexual reproduction. In general the numerical bounds on critical values from this method are terrible but Durrett (1992c) has shown that in some cases you can get good bounds.

Bramson and Neuhauser (1993) studied perturbations of one dimensional cellular automata. Their results are exciting because they apply to a number of examples that are not attractive. An important special case is that if one considers the addition mod 2 automaton:

$$
\eta_{n+1}(x) = (\eta_n(x-1) + \eta_n(x+1)) \pmod 2
$$

and adds spontaneous deaths at a small rate ϵ then there is a stationary distribution close to product measure with density 1/2. Figure 4.3 shows the cellular automaton starting from a single 1 at 0, which generates a discrete version of the Sierpinski gasket. Figure 4.4 shows what happens when we introduce spontaneous deaths at rate $\epsilon = 0.01$. Note that there are many more occupied sites in the model with extra deaths.

Figure 4.3. Pascal's triangle mod 2

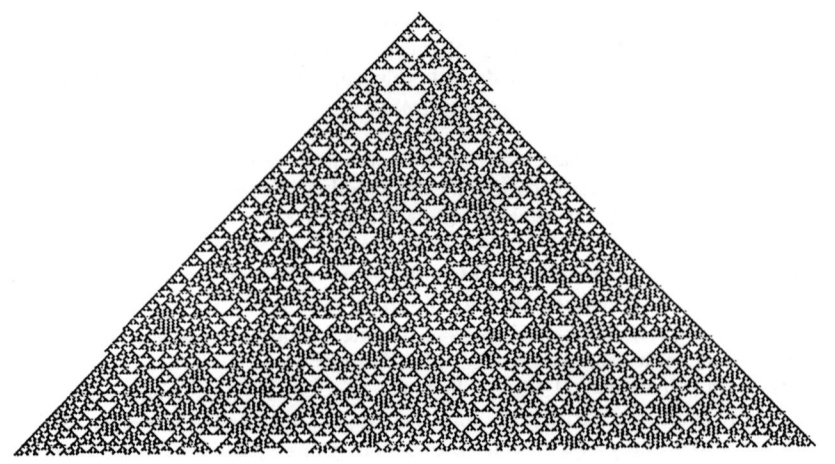

Figure 4.4. Plus spontaneous deaths with probability 0.01

5. Threshold Models

We begin by recalling a definition given in Section 1.

Example 5.1. The threshold voter model. The state space is $\{0,1\}^S$ and the flip rates are

$$c_i(x,\xi) = \begin{cases} 1 & \text{if } n_i(x,\xi) \geq \theta \\ 0 & \text{if } n_i(x,\xi) < \theta \end{cases}$$

Here, as usual, $n_i(x,\xi) = |\{y \in \mathcal{N} : \xi(x+y) = i\}|$ is the number of neighbors of type i and we assume $\mathcal{N} = \{y : \|y\|_p \leq r\}$ for some $1 \leq p \leq \infty$ and $r \geq 1$.

Our first goal is to show that the behavior of the threshold 1 voter model is much different from that of the basic voter model. We begin with one case in which the behavior is the same.

Theorem 5.1. Suppose $d = 1$ and $\mathcal{N} = \{-1,1\}$. Then the threshold 1 voter model clusters starting from any translation invariant initial state ξ_0. That is, for any $x \neq y$ we have $P(\xi_t(x) \neq \xi_t(y)) \to 0$.

PROOF: To motivate the proof, take a look at Figure 5.1 which shows a simulation of the system on $\{0, 1, \ldots, 719\}$ with periodic boundary conditions (i.e., 0 and 719 are neighbors). The initial configuration at the top of the page is product measure with density $1/2$. As we go down the page from time 0 at the top to time 690 at the bottom, it should be clear that intervals of sites with the same opinion can be destroyed but cannot be created. Thus the number of intervals per unit distance will go to 0, i.e., the system clusters.

To turn the last paragraph into a proof, we define a process on $1/2 + \mathbf{Z}$ so that

$$\zeta_t(x) = |\xi_t(x - 1/2) - \xi_t(x + 1/2)|$$

In words, there is a 1 at x if and only if $\xi_t(x - 1/2) \neq \xi_t(x + 1/2)$. ζ is called the *boundary process of* ξ since the 1's mark the boundaries between clusters of the voters with the same opinion. To see how ζ evolves consider the following picture

ξ	1		1		0		1		1		0		0
ζ		0		1		1		0		1		0	
x	1	1.5	2	2.5	3	3.5	4	4.5	5	5.5	6	6.5	7

Isolated 1's in ζ like the one at 5.5 perform random walks: the 1 at 5 flips to 0 at rate one and when this occurs the boundary jumps from 5.5 to 4.5; similarly, the 0 at 6 flips to 1 at rate one and when this occurs the boundary jumps from 5.5 to 6.5. When a 1 is adjacent to another 1 (like those at 2.5 and 3.5) they annihilate at rate 1, since when the 0 at 3 flips to 1 the two boundaries disappear.

Let $u(t) = P(\zeta_t(x) = 1)$, which is independent of x since we have supposed that ξ_0 is translation invariant. Since 1's can be destroyed in ζ but cannot be created, it should not be surprising that $u(t) \to 0$ as $t \to \infty$. To prove this, we note that

(5.1) $$\frac{du}{dt} = -P(\zeta_t(x) = 1, \zeta_t(x - 1) = 1) - P(\zeta_t(x) = 1, \zeta_t(x + 1) = 1)$$

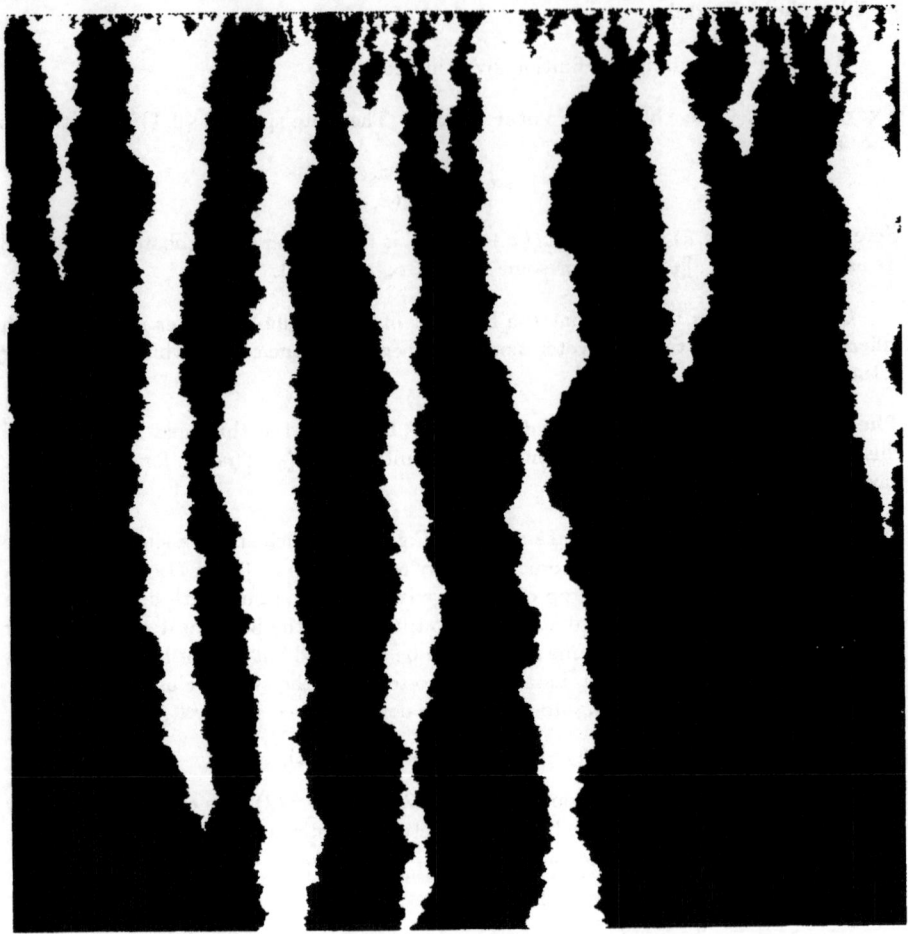

Figure 5.1. One dimensional nearest neighbor threshold voter model.

This can be proved by using $\frac{d}{dt}T_t f = T_t L f$ or more intuitively by noting that the right hand side gives the two ways that a 1 at x can be destroyed. The terms that involve a 1 moving to x or moving away from x cancel.

Translation invariance implies that the right hand side of (5.1) is

$$-2P(\zeta_t(0.5) = 1, \zeta_t(-0.5) = 1) \equiv -v(t)$$

The first step in proving $u(t) \to 0$ is to show that if $t \geq 1$

(5.2) $$v(t) \geq g(u(t-1)) \text{ where } g(x) > 0 \text{ when } x > 0$$

To do this we note that if $u(s) \geq 1/L$ where L is an integer then

(5.3) $$P(\zeta_s \text{ has at least two 1's in } (-L, L]) \geq \frac{1}{2L-1}$$

for otherwise we get a contradiction

$$2 \leq 2Lu(s) = E \sum_{x \in (-L,L]} \zeta_s(x) < 1 \cdot \frac{2L-2}{2L-1} + 2L \cdot \frac{1}{2L-1} = \frac{4L-2}{2L-1} = 2$$

Now if we have an initial configuration in which there are at least two 1's in $(-L, L]$ there is a probability $\geq \epsilon_L > 0$ that no particles will enter $(-L, L]$ before time 1, the two particles closest to 0 will move to 0.5 and -0.5, and none of the other particles in $(-L, L]$ will move. Combining this observation with (5.3) proves (5.2). To complete the proof of Theorem 5.1 now, we observe that $u(t)$ is decreasing so $u(t) \to u(\infty) \geq 0$ as $\to \infty$. If $u(\infty) > 0$ then for all t we have

$$\frac{du}{dt} = -v(t) \leq -g(u(\infty)) < 0$$

so integrating we find $u(t) \to -\infty$ a contradiction. $\qquad\qquad\qquad\square$

Remark. The argument above applies to any one dimensional nearest neighbor system in which $c_i(x, \xi) = f(n_i(x, \xi))$ with $f(0) = 0$, the so-called *nonlinear voter models*. In the case of the basic voter model, i.e., $f(2) = 2f(1)$ the boundary process is an *annihilating random walk*. That is, particles perform independent random walks until they hit at which time the two particles annihilate. Theorem 3.1 shows that for the basic voter model clustering occurs for any initial configuration. Theorem 4 in Cox and Durrett shows that for the threshold voter model clustering occurs for any initial configuration. We

Conjecture 5.1. In any one dimensional nearest neighbor nonlinear voter model clustering occurs for any initial configuration.

Our next goal is to show that coexistence is possible in the threshold 1 voter model even in one dimension. To do this we will use some ideas from Liggett (1993) to compare with

Example 5.2. The threshold contact process. The state space is $\{0,1\}^S$ and the flip rates are

$$c_1(x,\xi) = \begin{cases} \lambda & \text{if } n_1(x,\xi) \geq \theta \\ 0 & \text{if } n_1(x,\xi) < \theta \end{cases}$$

$$c_0(x,\xi) = 1$$

Here $c_1(x,\xi)$ is the same as in the threshold voter model but we have set $c_0(x,\xi) \equiv 1$.

(5.4) Lemma. If the threshold θ contact process with $\lambda = 1$ has a nontrivial stationary distribution then so does the threshold θ voter model.

PROOF: To construct the stationary distribution we will start the threshold voter model ξ from $\nu_{1/2}$, product measure with density $1/2$, and compare with the threshold contact process ζ to show that clustering does not occur.

The first step in doing this is to show that the upper invariant measure π for the threshold voter model with $\lambda = 1$ is stochastically smaller than $\nu_{1/2}$ To do this we compare the threshold contact process ζ with the "independent flips process" η_t in which $c_i(x,\eta) \equiv 1$, i.e., each site flips at rate 1 independently of the others. Since sites in η flip to 1 at rate one independent of what is around them, if we start ζ and η with $\zeta_0 = \eta_0$ having distribution π and construct the two processes using the recipe in Section 2 then $\zeta_t(x) \leq \eta_t(x)$ for all t and x. This is true since 1's flip to 0 at rate 1 in both processes while 0's flip to 1 at rate 1 always in η, but at rate 1 in ζ only if there are enough 1 neighbors. On the graphical representation then we find that each flip preserves the inequality and the result can be proved like (2.5).

Now since the sites in η flip independently it is easy to see that as $t \to \infty$ η_t converges to $\nu_{1/2}$. The inequality $\zeta_t(x) \leq \eta_t(x)$ and the fact that ζ_t always has distribution π imply that π is stochastically smaller than $\nu_{1/2}$. To prove this we observe that if f is increasing and depends on only finitely many coordinates then $Ef(\zeta_t) \leq Ef(\eta_t)$ and since any such f is bounded and continuous letting $t \to \infty$ gives

$$\int f(\xi)d\pi \leq \int f(\xi)d\pi$$

checking the definition we gave in (2.11).

Now the result of Holley in the remark (2.12) implies that we can define ξ_0 with distribution $\nu_{1/2}$ and ζ_0 with distribution π, so that that $\xi_0(x) \geq \zeta_0(x)$ for all x. Since sites in ζ flip to 0 at rate one, while those in ξ only flip to 0 at rate one when there are enough 0 neighbors, and the rates of flipping to 1 are the same, if we construct the two processes using the recipe in Section 2 then $\xi_t(x) \geq \zeta_t(x)$ for all x and t. To construct a stationary distribution for ξ, let μ_t be the distribution of ξ_t, form the Cesaro average

$$\bar{\mu}_T = \frac{1}{T} \int_0^T \mu_t dt$$

and let $\bar{\mu}_\infty$ be the limit of a weakly convergent subsequence. It follows from (2.13) that $\bar{\mu}_\infty$ is a stationary distribution. To see that it concentrates on configurations with infinitely many 1's we note that the inequality $\xi_t(x) \geq \zeta_t(x)$ implies that $\bar{\mu}_\infty$ is larger than the upper invariant measure π, which is spatially ergodic by (2.15) and hence concentrates on configurations with infinitely many 1's. To see that $\bar{\mu}_\infty$ concentrates on configurations with infinitely many 0's, note that the initial distribution $\nu_{1/2}$ and the threshold voter model are symmetric under the interchange of 0's and 1's, so the limit measure $\bar{\mu}_\infty$ is as well. □

Liggett (1993) has shown

Theorem 5.2. When $d = 1$ and $\mathcal{N} = \{-2, -1, 1, 2\}$ or $d = 2$ and $\mathcal{N} = \{y : \|y\|_1 = 1\}$ the threshold 1 contact process with $\lambda = 1$ has a nontrivial stationary distribution.

Since enlarging the neighborhood $\mathcal{N}$ makes it easier for the threshold 1 contact process to have a nontrivial stationary distribution, it follows from (5.4) and Theorem 5.2 that

Theorem 5.3. Suppose $\mathcal{N} = \{z : \|z\|_p \leq r\}$ with $1 \leq p \leq \infty$ and $r \geq 1$. With the exception of the one dimensional nearest neighbor case, the threshold one voter model always has a nontrivial stationary distribution.

By another comparison argument Liggett shows that to prove Theorem 5.3 it is enough to consider the case $d = 1$ and $\mathcal{N} = \{-2, -1, 1, 2\}$ - map $\mathbf{Z}^2$ to $\mathbf{Z}$ by $(x, y) \rightarrow x + 2y$ and notice that the image of the two dimensional threshold contact process dominates the one dimensional one. A simulation of the case $d = 1$ and $\mathcal{N} = \{-2, -1, 1, 2\}$ given in Figure 5.2, which parallels the one for the nearest neighbor case in Figure 5.1, makes it clear that Theorem 5.3 is true. However, the proof of Theorem 5.2 (which implies 5.3) requires a tricky generalization of the result Holley and Liggett (1978) that the one dimensional nearest neighbor contact process has $\lambda_c \leq 2$. Therefore we content ourselves to prove less (and more).

Theorem 5.4. Suppose $\mathcal{N} = \{y : \|y\|_p \leq r\}$ with $r \geq 1$. For any threshold θ if $r \geq r_0(d, \theta)$ then there is a nontrivial stationary distribution for threshold θ contact process with $\lambda = 1$ and hence also for the threshold θ voter model.

PROOF: We will use the comparison theorem from Section 4. To do this, it is convenient to suppose that ξ has been constructed from a percolation substructure with rate 1 Poisson processes $\{T_n^x, n \geq 1\}$ at which times we draw arrows from $y + x$ to x for all $y \in \mathcal{N}$, and rate 1 Poisson processes $\{U_n^x, n \geq 1\}$ at which times we write a δ at x.

Exercise. This shows that the threshold contact process can be constructed from a percolation substructure defined in Section 3. What is the dual process?

Suppose $r = (2d + 2)L$. To check the comparison assumptions, let H be the configurations that have at least θ 1's in $[-L, L]^d$. Let $\gamma > 0$. If T is small enough then the probability that $U_1^x > T$ for all of our θ 1's, is $e^{-\theta T} > 1 - \gamma/5$. Now since $r = (2d + 2)L$, the neighborhood of each site in $I_1 = [L, 3L] \times [-L, L]^{d-1}$ contains all the sites in $[-L, L]^d$

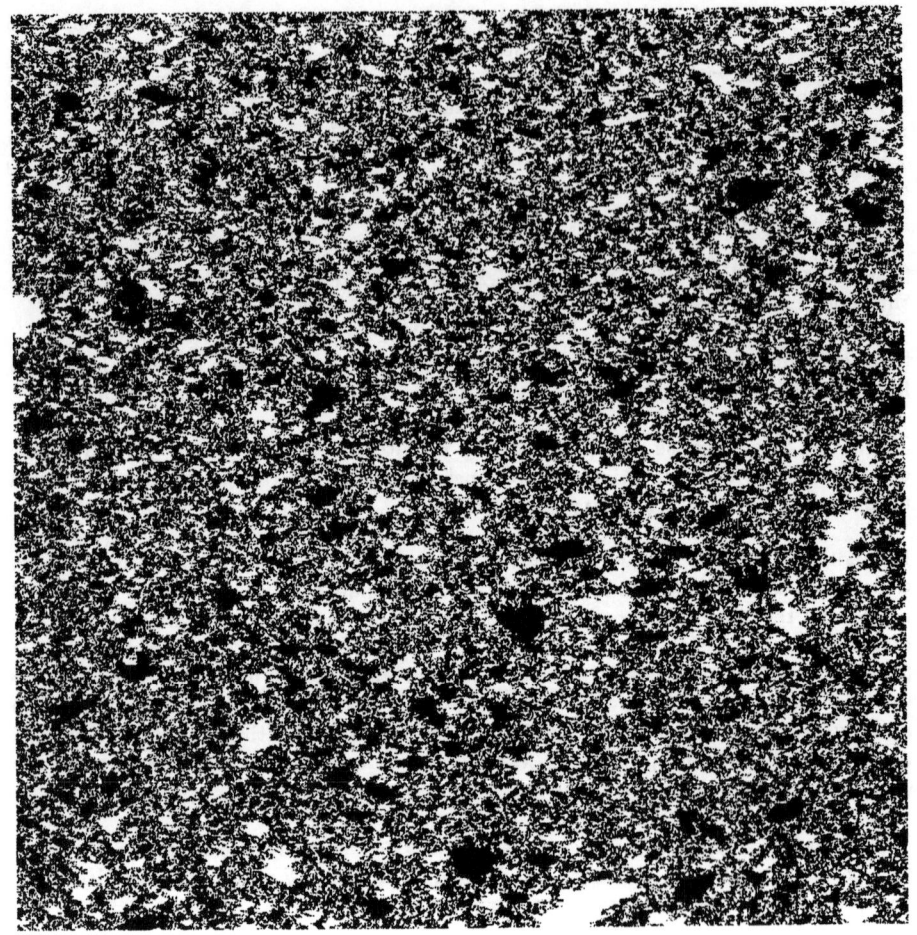

Figure 5.2. One dimensional threshold voter model, range two.

(distances are largest for the L^1 norm and for the points $(3L, L, \ldots, L)$ and $(-L, L, \ldots, L)$).
Now as long as there are at least θ 1's in $[-L, L]^d$, each site in $[L, 3L]$ will flip to 1 at rate 1.
If r and hence L is sufficiently large then with probability at least $1 - \gamma/5$ at least θ sites will
flip to 1 by time T. A similar remark applies to the sites in $I_{-1} = [-3L, -L] \times [-L, L]^{d-1}$,
and our first estimate implies that in each case the probability one of our θ 1's flips back
to 0 by time T is $\leq \gamma/5$.

The results in the last paragraph show that if we start with θ 1's in $I_0 = [-L, L]^d$
then with probability at least $1 - \gamma$ there will be at least θ 1's in I_1 and in I_{-1} at time T.
Our good event is measurable with respect to the graphical representation in $[-3L, 3L]^d$
so we have checked the comparison assumptions of Section 4 with $k_0 = 3$ and $j_0 = 1$. If
we start the threshold contact process with all sites occupied then Theorem 4.3 implies
our process dominates an oriented percolation starting with all sites wet, so Theorem 4.2
shows

$$\liminf_{n \to \infty} P(0 \in X_{2n}) \geq 19/20$$

Now $0 \in X_{2n}$ means that there are at least θ 1's in $[-L, L]^d$ at time $2nT$ and ξ_{2nT} is
translation invariant so it follows that

$$\liminf_{n \to \infty} P(\xi_{2nT}(0) = 1) = \liminf_{n \to \infty} \frac{1}{(2L+1)^d} \sum_{x \in [-L,L]^d} P(\xi_{2nT}(x) = 1)$$

$$\geq \frac{1}{(2L+1)^d} \cdot \theta \cdot \frac{19}{20} > 0$$

To pass from this result to the whole sequence we notice that since a 1 survives for t units
of time with probability e^{-t}, $P(\xi_{2nT+t}(0) = 1) \geq e^{-t} P(\xi_{2nT}(0) = 1)$. Combined with the
last result this implies

$$\liminf_{T \to \infty} \frac{1}{T} \int_0^T P(\xi_s(0) = 1)\, ds > 0$$

and it follows from (2.13) that there is a nontrivial stationary distribution. □

The last result shows that if the threshold is small compared to the number of neighbors then *coexistence* occurs in the threshold voter model, i.e. there is a stationary distribution that concentrates on configurations with infinitely 1's and infinitely many 0's.
Our next result due to Durrett and Steif (1993) shows that if the threshold is too large
the system *fixates*, i.e., with probability one each site changes its state only finitely many
times.

Theorem 5.5. Suppose $\mathcal{N} = \{y : \|y\|_p \leq r\}$. If $\theta > (|\mathcal{N}| - 1)/2$ then the system fixates.

The borderline case in this result, $\theta = (|\mathcal{N}|+1)/2$ ($|\mathcal{N}|$ is always odd), is called *the majority
vote process*, since you change your mind if you are in the minority in your neighborhood.

PROOF: Our proof is based on an idea of Grannan and Swindle. Let $\delta_{x,y}(t)$ be 1 if
$\xi_t(x) \neq \xi_t(y)$, 0 otherwise, and define the *energy* at time t to be

$$\mathcal{E}_t = \sum_{x,y:y-x \in \mathcal{N}} e^{-\epsilon \|x+y\|_2} \delta_{x,y}(t)$$

where $\epsilon > 0$ is to be chosen later. Since $0 \leq \mathcal{E}_0 < \infty$, we can prove Theorem 5.5 by showing

(5.5) If $\theta > (|\mathcal{N}| - 1)/2$ and ϵ is small then a flip at x decreases the energy by at least $\gamma(x) > 0$.

To prove (5.5) we note that if $\alpha = |\{y \in x + \mathcal{N} : \xi_t(y) \neq \xi_t(x)\}|$ and $N = \sup\{\|x\|_2 : x \in \mathcal{N}\}$ then the drop in energy due to a flip at x is at least

$$(5.6) \qquad e^{-2\epsilon\|x\|_2} \left[e^{-\epsilon N}\alpha - e^{\epsilon N}(|\mathcal{N}| - 1 - \alpha) \right]$$

since (i) the site x now agrees with the α sites it used to disagree with and now disagrees with the other $|\mathcal{N}| - 1 - \alpha$ neighbors and (ii) even in the worst case all the points in $\{y \in x + \mathcal{N} : \xi_t(y) \neq \xi_t(x)\}$ have $\|x + y\|_2 \leq 2\|x\|_2 + N$ and the other points $y \in x + \mathcal{N}$ have $\|x + y\|_2 \geq 2\|x\|_2 - N$. In order for a flip to occur we must have $\alpha \geq \theta > (|\mathcal{N}| - 1)/2$ and hence $|\mathcal{N}| - 1 - \alpha < \alpha$. Since the last two number are integers smaller than $|\mathcal{N}|$, (5.5) follows from (5.6). $\qquad\square$

Refinements of Theorem 5.4. Before we stated Theorem 5.5, we said "if the threshold is small compared to the number of neighbors" then the threshold contact process with $\lambda = 1$ has a nontrivial stationary distribution (and hence there is coexistence in the threshold voter model). What we would like to concentrate on now is:

How large can θ be when the range is r?

The comparison theorem involves obnoxiously small constants (when $M = 1$ Theorems 4.1 and 4.2 require $\gamma \leq 6^{-100}$). So we cannot hope to get a nontrivial result for $r = 10$, or even $r = 10,000$, but it is not unreasonable to look at how θ behaves asymptotically with r. The results were are about to give foreshadow the developments in the next section, but are not needed for them, or for any subsequent section, and can be skipped without loss.

Here and until the end of the section we suppose $\mathcal{N} = \{z : \|z\|_p \leq r\}$, let $N = |\mathcal{N}|$, and we investigate what happens for fixed p as $r \to \infty$ First let's see what we get when we follow the proof of Theorem 5.4.

(5.7) There is a $c_p > 0$ so that if $\theta \leq c_p\sqrt{N}$ and if r (and hence N) is large then the threshold θ contact process with $\lambda = 1$ has a nontrivial translation invariant stationary distribution.

PROOF: Taking $T = \gamma/5\theta$ gives $e^{-\theta T} = e^{-\gamma/5} \geq 1 - \gamma/5$. Having fixed the time, the number of sites in $[L, 3L] \times [-L, L]^{d-1}$ that flip to 1 by time T has a binomial distribution with parameters $n = (2L + 1)^d$ and $p = 1 - e^{-T} \geq \gamma/60$ when θ is large. If we let Z be the number of sites in $[L, 3L] \times [-L, L]^{d-1}$ that flip to 1 by time T then Z has mean $\geq (2L+1)^d\gamma/60$ and variance $\leq (2L+1)^d\gamma/60$ so if we set $(2L+1)^d\gamma/60 = 2\theta$ (sticklers for details should take the smallest integer L so that $\geq$ holds) Chebyshev's inequality implies that

$$P(Z \leq \theta) \leq \frac{(2L+1)^d\gamma/60}{\theta^2} \leq \frac{2}{\theta} \to 0$$

as $\theta \to \infty$. Now $\theta^2 = (2L+1)^d \gamma/12 \geq c_p N$ since $r = (2d+2)L$ and the result follows. $\square$

By choosing a more intelligent block event we can get

(5.8) There is a $c_p > 0$ so that if $\theta \leq c_p N$ and if r (and hence N) is large then the threshold θ contact process with $\lambda = 1$ has a nontrivial translation invariant stationary distribution.

PROOF: Let $\theta = (2L+1)^d/5$ and let H be the configurations that have at least $(2L+1)^d/4$ 1's in $[-L, L]^d$. If we pick $r = (2d+2)L$ then $\theta \geq c_p N$ for all r and as long as there are at least θ 1's in $[-L, L]^d$ the number of 1's in $[-L, L]^d$ (or in $[L, 3L] \times [-L, L]^{d-1}$), behaves like a Markov chain that jumps $k \to k+1$ at rate $(2L+1)^d - k$ and $k \to k-1$ at rate k. Now when $k \leq (2L+1)^d/3$ this chain jumps at rate $(2L+1)^d$ moving up with probability at least $2/3$ and down with probability at most $1/3$. A comparison with asymmetric simple random walk shows

(i) with high probability it will take a long time (i.e., at least $e^{c(2L+1)^d}$ for some $c > 0$) for the total number of 1's in $[-L, L]^d$ to go below θ

(ii) we can pick a large time T (that is independent of L) so that if L is large then with high probability the number of 1's in $[L, 3L] \times [-L, L]^{d-1}$ and in $[-3L, -L] \times [-L, L]^{d-1}$ at time T will be at least $(2L+1)^d/4$

We leave it to the reader to fill in the missing details since we know how to prove a sharp result:

(5.9) Let $c < 1/4$. If $\theta \leq cN$ and if r (and hence N) is large then the threshold θ contact process with $\lambda = 1$ has a nontrivial translation invariant stationary distribution. Let $c > 1/4$. If $\theta \geq cN$ and if r (and hence N) is large then the threshold θ contact process with $\lambda = 1$ has only the trivial stationary distribution.

The proof of the first conclusion is closely related to that of Theorem 6.1. For details and the proof of the converse see Durrett (1992).

6. Cyclic Models

As already suggested by our remarks on refinements in the last section, we can considerably close the gap between Theorems 5.4 and 5.5 if we look at systems with large range. The proof of our main result, Theorem 6.1, is no harder for a class of models that includes a multicolor version of the threshold voter model, so we formulate the result in that generality.

Example 6.1. Cyclic Color Model. The states of each site are $\{0, 1, \ldots, \kappa - 1\}$ and the flip rates are

$$c_i(x, \xi) = \begin{cases} 1 & \text{if } \xi(x) = i - 1 \text{ and } n_i(x, \xi) \geq \theta \\ 0 & \text{otherwise} \end{cases}$$

Here and throughout this section, arithemtic is done modulo κ so $0 - 1 = \kappa - 1$. When $\kappa = 2$ the last definition reduces to the threshold voter model. The dynamics here were invented by David Griffeath as a generalization of the voter model. The cyclic color model is closely related to the hypercycle of evolutionary biology. See Eigen and Schuster (1979) and Boerlijst and Hogeweg (1991).

Our main result also applies to two other examples

Example 6.2. Greenberg Hastings Model. The states of each site are $\{0, 1, \ldots, \kappa - 1\}$ and the flip rates are

$$\begin{aligned} c_1(x, \xi) &= 1 & \text{if } \xi(x) = 0 \text{ and } n_1(x, \xi) \geq \theta \\ c_i(x, \xi) &= 1 & \text{if } \xi(x) = i - 1 \end{aligned}$$

In words, we need an above threshold number of 1's to make the transition from $0 \to 1$ but then the rest of the transitions happen at rate 1. When $\kappa = 2$ this reduces to the threshold contact process with $\lambda = 1$.

Example 6.3. Host Parasitoid Interactions. Insect parasitoids lay their eggs on or in the bodies of other arthropods, and the parasitoid larvae kill their host as they feed on it. Hassell, Comins, and May (1991) introduced a cellular automaton model for this system. The corresponding particle system model has nine states $\{0, 1, \ldots 8\}$ and makes transitions as follows:

$$\begin{aligned} c_1(x, \xi) &= 1 & \text{if } \xi(x) = 0 \text{ and } n_1(x, \xi) \geq \theta \\ c_4(x, \xi) &= 1 & \text{if } \xi(x) = 3 \text{ and } n_5(x, \xi) \geq \theta \\ c_i(x, \xi) &= 1 & \text{if } i \neq 1, 4 \text{ and } \xi(x) = i - 1 \end{aligned}$$

As they explain on page 256, the first transition corresponds to colonization of empty sites (state 0) by the host, the second to a mature parasitoid (state 5) colonizing a mature host (state 3), and the others to the aging and/or death of host and parasitoid.

To indicate what common features of the last three models are needed to apply Theorem 6.1, we say that ξ is a *cyclic model* if the states of each site are $\{0, 1, \ldots, \kappa - 1\}$ and makes transitions as follows:

$$c_i(x, \xi) = 1 \qquad \text{if } \xi(x) = i \text{ and } n_{g(i)}(x, \xi) \geq \theta_i$$

Here $g(i) \in \{0, 1, \ldots, \kappa - 1\}$ and we set $\theta_i = 0$ if the transition happens at rate 1 independent of the states of the neighbors. Let $\theta = \max_i \theta_i$.

Theorem 6.1. Let $\epsilon > 0$ and suppose $\theta \leq (1 - \epsilon)|\mathcal{N}|/2\kappa$. If $r \geq R_\epsilon$ then there is a stationary distribution close to the uniform product measure.

Recall that we suppose $\mathcal{N} = \{y : \|y\|_p \leq r\}$ and that the uniform product measure is the one in which the coordinates are independent and have $P(\xi(x) = i) = 1/\kappa$. When $\kappa = 2$ this says that for thresholds $a|\mathcal{N}|$ with $a < 1/4$ there is coexistence for large r. (This result was stated in (5.9).) In contrast Theorem 5.4 says that when $a \geq 1/2$ the system fixates for any r. We

Conjecture 6.1. When $\theta = a|\mathcal{N}|$ in the threshold voter model and $1/4 < a < 1/2$, clustering occurs for large r.

We will explain our reasons after we give the proof. Theorem 6.1 concentrates on the behavior for large range. For results about the one dimensional cyclic color model, see Bramson and Griffeath (1987) (1989), or for a treatment of the corresponding cellular automaton, see Fisch (1990a), (1990b), (1991).

PROOF IN $d = 1$: Let $a = \theta/|\mathcal{N}|$. By assumption $a \leq (1 - \epsilon)/2\kappa$. Pick $\beta \in (0, \epsilon/4]$ so that $B = 1/\beta$ is an integer, pick $\rho < \sigma < 1/\kappa$ so that $(1 - 2\beta)\rho \geq (1 - \epsilon)/\kappa$, then pick r large enough so that

$$(1 - \beta)\rho \cdot \frac{2r}{2r + 1} \geq (1 - \epsilon)/\kappa$$

Let $K = \beta r$ and note that $BK = r$. For each $m \in \mathbf{Z}$, we call $[mK, (m + 1)K)$ a *house*. We say that a house is *good* at time 0 if it contains at least σK sites in each of the states $0, 1, \ldots, \kappa - 1$. We say that the interval $[-r, r)$ is *good* at time 0 if all the houses it contains are good. This will be our event H when we apply Theorem 4.3.

We have chosen our constants so that as long as each house in $[-r, r)$ is *reasonable* i.e., contains at least ρK sites of each color, each site in $[-r - K, r + K)$ will see at least θ sites of each color. To check this, note that the worst case occurs when $x \in [r, r + K)$, but even in this case all the sites in $[K, r)$ are in its neighborhood and if all of the houses in $[K, r)$ are reasonable then the number of sites of a given color in x's neighborhood will be at least

$$\rho(r - K) = \rho r(1 - \beta) = \frac{\rho(1 - \beta)}{2} \cdot \frac{2r}{2r + 1} \cdot 2r + 1$$

$$\geq \frac{(1 - \epsilon)}{2\kappa} \cdot (2r + 1) \geq \theta$$

So as long as each house in $[-r, r)$ stays reasonable, the sites in $[-r - K, r + K)$ flip from i to $i + 1$ at rate 1 (here $(\kappa - 1) + 1 = 0$) and hence behave like independent Markov chains. These "single site" Markov chains are irreducible on $\{0, 1, \ldots, \kappa - 1\}$ and hence converge to the equilibrium distribution, which assigns probability $1/\kappa$ to each state. Let $p_t(i, j)$ be the transition probability of the single site Markov chain, let $\sigma' \in (\sigma, 1/\kappa)$ and pick S so that $p_S(0, i) \geq \sigma'$ for all i. Let $T = 2BS$. By using a simple large deviations result (see (6.2) below) it is easy to show that with high probability

(a) All the houses in $[-r, r)$ stay reasonable until time T.

(b) The houses $[r + (j - 1)K, r + jK)$ and $[-r - jK, -r - (j - 1)K)$ will be good at time jS and stay reasonable to time T.

(c) All the houses in $[r, 3r)$ and $[-3r, -r)$ will be good at time T.

Figure 6.1 gives a picture of this expansion. The gray shaded area gives the space time region occupied by reasonable houses.

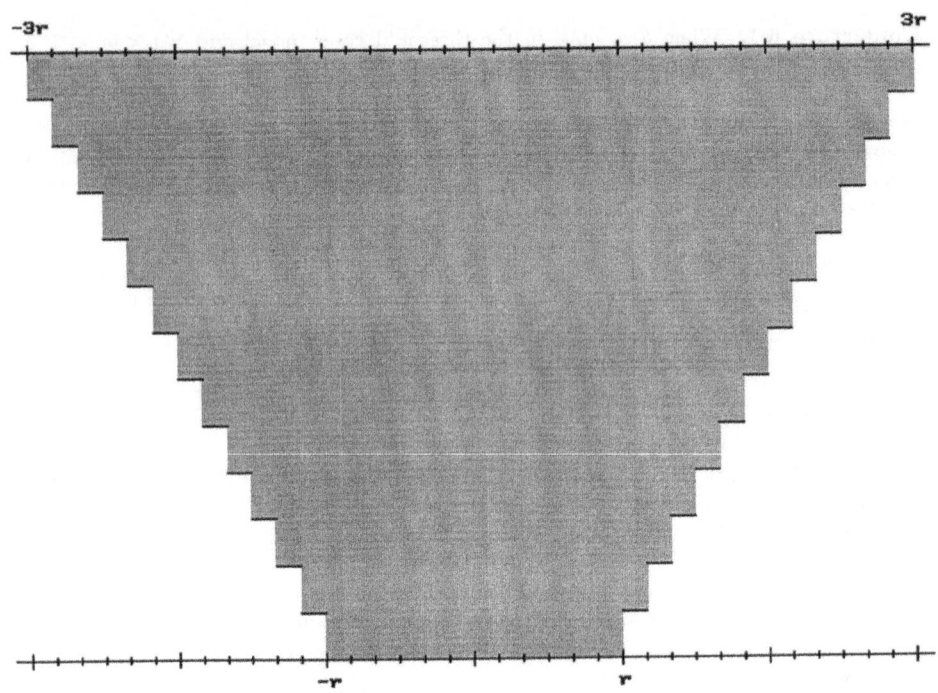

Figure 6.1

Since our good event is measurable with respect to the Poisson processes in $[-3r, 3r) \times [0, T)$ we have verified the comparison assumptions with $L = r$, $K = 3$, $J = 1$. If we start our cyclic system from uniform product measure then X_0 is a Bernoulli set with density $p > 0$. (p is close to 1 if L is large but we do not need that.) Applying Theorems 4.3 and 4.2 now it follows that

$$\liminf_{n \to \infty} P(0 \in X_n) \geq 19/20$$

Arguing as in the end of the proof of Theorem 5.4 it is easy to improve this conclusion to

$$\liminf_{n \to \infty} P(\text{ all } \kappa \text{ colors are in } [-r,r)) > 0$$

and it follows from (2.13) that there is anontrivial stationary distribution. By using an improvement of Theorem 4.2 given in the appendix (see Theorem A.3)

(6.1) **Lemma.** If $p > 0$ and $\gamma \leq 6^{-4(2M+1)^2}$ then

$$\liminf_{n \to \infty} P(\{-2K,\ldots,2K\} \cap W_{2n}^p \neq \emptyset) \geq 1 - \epsilon_K$$

where $\epsilon_K \to 0$ as $K \to \infty$

we can show that the stationary distribution we constructed concentrates on configurations in which there are infinitely many sites in each state. ((6.1) shows directly that with probability one each state appears somewhere in the configuration, but the distribution is stationary so if there were only finitely many sites in some state we would have positive probability of having 0 in that state a contradiction.)

By the arguments in the last paragraph it is enough to show that (a), (b), and (c) hold. The first step is proving the large deviations estimate.

(6.2) **Lemma.** Let $X_1,\ldots X_n$ be i.i.d. with $P(X_i = 1) = p$, $P(X_i = 0) = 1 - p$. Then

$$P(X_1 + \ldots + X_n \leq n(p - \epsilon)) \leq \exp(-\epsilon^2 n/2).$$

Remark. This result and its proof are standard but we need to know that the right hand side does not depend on p.

Proof: If $\alpha > 0$ then

$$P(X_1 + \ldots + X_n \leq n(p - \epsilon)) e^{-\alpha n(p-\epsilon)} \leq (pe^{-\alpha} + (1 - p))^n$$

Taking log's, dividing by n, rearranging and then using $\log(1 + x) \leq x$ we have

$$\frac{1}{n} \log P(X_1 + \ldots + X_n \leq n(p - \epsilon)) \leq \alpha(p - \epsilon) + \log(1 + p(e^{-\alpha} - 1))$$

$$\leq \alpha(p - \epsilon) + p(e^{-\alpha} - 1) = -\alpha\epsilon + p(e^{-\alpha} - 1 + \alpha)$$

Now $e^{-\alpha} - 1 + \alpha = \alpha^2/2! - \alpha^3/3! + \ldots \leq \alpha^2/2$ for $0 < \alpha < 1$, so taking $\alpha = \epsilon$ and using $p \leq 1$ gives

$$P(X_1 + \ldots + X_n \leq n(p - \epsilon)) \leq \exp(-\epsilon^2 n/2)$$

and completes the proof of (6.2). $\square$

Let Z_t be a copy of the single site Markov chain, let $p_t(i, j) = P_i(Z_t = j)$, and observe that $p_t(i, j) = p_t(0, j - i)$. Until the first time some house in $[-r, r)$ becomes unreasonable, the sites in each house in $[-r,r)$ behave like independent copies of the single site Markov chain so we consider a collection of $K = r\beta$ independent copies of Z_t and let v_i be the number of "sites" in state i at time 0. The expected number of sites in state j at time t is $w_j(t) = \sum_i v_i p_t(i,j)$. To prove (a) we apply (6.2) with $n = v_i \geq \sigma K$ to the sites that start in state i to see that with probability at least $1 - \exp(-\epsilon^2 \sigma K/2)$, at least $v_i(p_t(i,j) - \epsilon)$ of the sites that start in state i will be in state j at time t. Taking $\epsilon = (\sigma - \rho)$ and summing over i gives

$$\sum_i v_i(p_t(i,j) - \epsilon) \geq \sigma K \sum_i p_t(i,j) - K\epsilon \geq (\sigma - \epsilon)K = \rho K$$

since $\sum_i v_i = K$ and $\sum_i p_t(i,j) = \sum_i p_t(0, j - i) = 1$. So with probability at least $1 - \kappa \exp(-\epsilon^2 \sigma K/2)$, at least ρK sites will be in state j at time t.

The last bound is for a fixed time but it is easy to extend it to cover the interval $[0,T]$. Let $\delta = \epsilon^2 \sigma / 2$, let $J = \exp(\delta K/2)$, and $t_k = k/J$ for $1 \leq k \leq JT$. The probability that the number of sites in state i is less than ρK at some time t_k is at most

$$\kappa JT \exp(-\epsilon^2 \sigma K/2) = \kappa T \exp(-\delta K/2)$$

The probability that two sites flip in some interval (t_{k-1}, t_k) is at most

$$JT \binom{K}{2} J^{-2} \leq K^3 T \exp(-\delta K/2).$$

When we never have two flips in any interval, the state at each $t \in (t_{k-1}, t_k)$ agrees with the state at one of the two endpoints. Combining the last two estimates we have that the probability a collection of K independent single site chains becomes unreasonable before time T

$$\leq (\kappa + K^3)T \exp(-\delta K/2)$$

Since the sites in $[-r,r)$ behave like independent single site chains until some house becomes unreasonable, the probability of the event in (a) is at least

$$1 - 2B(\kappa + K^3)T \exp(-\delta K/2)$$

The proof that the house $[r, r + K)$ will be good at time S is similar but simpler. If all the houses in $[-r,r)$ stay reasonable until time S then each site in $[r, r + K)$ always sees an above threshold number of sites of each color and flips to the next color at rate 1. We again consider a collection of K independent single site chains but this time starting from an arbitrary initial configuration. The choice of S guarantees that $p_S(i,j) \geq \sigma'$ so applying (6.2) to K i.i.d. random variables with $p = \sigma'$ we conclude that the fraction of sites in state j is at least σK with probability at least $1 - \kappa \exp(-(\sigma' - \sigma)^2 K/2)$. Once we know that with high probability $[r, r + K)$ is good at time S and all the houses in $[-r,r)$ are reasonable at all times in $[0,T]$, we can repeat the proof of (a) to conclude that the house $[r, r + K)$ stays reasonable at all times in $[S,T]$. This verifies (b) when $j = 1$ but by continuing in the same way we can prove the result for $2 \leq j \leq 2B$. Now (b) implies that all the houses in $[-3r + K, 3r - K)$ are reasonable at time $T - S$ we can repeat the proof that the house $[r, r + K)$ is good at time S to conclude that all the houses in $[-3r, 3r)$ are good at time T and the proof is complete. □

PROOF IN $d > 1$: Let $B_p(x, r) = \{y : \|x - y\|_p \leq r\}$. The key to the proof is the following fact, which basically says that large balls are almost flat.

(6.3) **Lemma.** *Suppose* $\lambda < 1/2$. *There are constants* R_0, δ, *and* M_0, *so that if* $M \geq M_0$ *and* $R \geq R_0$ *then for* $x \in B_2(0, (R + \delta)M)$.

$$|B_2(0, RM) \cap B_p(x, M)| \geq \lambda |B_p(x, M)|$$

PROOF: In one dimension we can take $R_0 = 1$ and $\delta = 1 - 2\lambda$. Turning to dimensions $d > 1$, let $Q = \{x \in R^d : \|x\|_p \leq 1\}$ and let q be its volume. To prove the result it is convenient to scale space by $1/M$ and translate so that x/M sits at the origin. Any $d - 1$ dimensional hyperplane through the origin divides Q into two pieces with volume $q/2$. For $i = 1, 2, 3$ let $\lambda < \lambda_3 < \lambda_2 < \lambda_1 < 1/2$. By continuity, there is a $\delta > 0$ so that if a hyperplane passes within a distance δ of the origin then it divides Q into two pieces each of which has volume at least $q\lambda_1$. Another application of continuity shows that if R_0 is large and $D = B_2(y, r)$ with $r \geq R_0$ and $B_2(y, r) \cap B_2(0, \Delta) \neq \emptyset$ then the volume of $D \cap Q$ is at least $q\lambda_2$.

The last step is to argue that if M is large then the lattice behaves like the "continuum limit" considered above. Pick $\epsilon > 0$ so that if $D = B_2(y, r)$ is as above then the volume of $B_2(y, r - \epsilon) \cap (1 - \epsilon)Q$ is always larger than $q\lambda_3$. Then pick M_0 so that $1/M_0 < \epsilon$ and if $M \geq M_0$ then $|B_p(0, M)|/qM^d < \lambda_3/\lambda$. Let $\mathcal{X} = (Z^d/M) \cap D \cap Q$. The first part of the choice of M_0 implies that if $M \geq M_0$ then

$$B_2(y, r - \epsilon) \cap Q(1 - \epsilon) \subset \cup_{x \in \mathcal{X}} x + \left[\frac{-1}{2M}, \frac{1}{2M}\right]^d$$

so $M^{-d}|\mathcal{X}| \geq q\lambda_3 \geq \lambda |B_p(0, M)| M^{-d}$, by the second part of the choice of M_0 and the proof is complete. □

To use this lemma we pick $\lambda < 1/2$ and $\rho < 1/\kappa$ so that $\lambda\rho > a$, use (6.3) to pick R_0, Δ, M_0, and then pick $M_1 \geq M_0$ so that

(6.4) $\qquad \lambda\rho K^d |B_p(0, M_1)| > a|B_p(0, K(M_1 + d))|$ holds for large K

Let $\sigma \in (\rho, 1/\kappa)$ and suppose that the range of interaction is $r = K(M_1 + d)$. For $z \in \mathbf{Z}^d$ let

$$I_z = [z_1 K, (z_1 + 1)K) \times \cdots \times [z_d K, (z_d + 1)K)$$

and call I_z a *house*. We say that a house is *good* at time 0 if it contains at least σK^d sites of each color. We say that ξ_0 is good if all the houses I_z, $z \in B_2(0, R_0 M_1)$ are good. This will be our event H when we apply Theorem 4.3.

We have set things up so that as long as each house in $B_2(0, R_0 M_1)$ is *reasonable*, i.e. contains at least ρK^d sites of each color, each site in each house in $B_2(0, (R_0 + \delta)M_1)$ sees at least θ sites of each color. To check this note if $z \in B_2(0, (R_0 + \delta)M_1)$ then all the sites in any house I_w with $w \in B_2(0, R_0 M_1) \cap B_p(z, M_1)$ are within p-norm distance $r = (M_1 + d)K$ of each site in I_z. (To see note that $\|z - w\|_p \le M_1$ so $\|zK - wK\|_p \le M_1 K$ and if we use 1 to denote a vector of 1's then $\|zK - (w + 1)K\|_p \le (M_1 + d)K$, with $p = 1$ being the worst case.) By (6.3)

$$|B_2(0, R_0 M_1) \cap B_p(x, M_1)| \ge \lambda |B_p(x, M_1)|$$

Multiplying the last inequality by ρK^d and using the choice of M_1 and K in (6.4) that

$$\rho K^d |B_2(0, R_0 M_1) \cap B_p(x, M_1)| \ge \lambda \rho K^d |B_p(x, M)|$$
$$\ge a|B_p(x, K(M_1 + d)| = \theta$$

Pick B so that $B\delta > 2R_0$ and hence

$$B_p(x, (R_0 + B\delta)M_1) \supset B_p(x, 3R_0 M_1)$$

Let $\sigma' \in (\sigma, 1/\kappa)$, choose S so that $p_S(0, i) \ge \sigma'$ for all i, and let $T = BS$. Let $D_j = B_2(0, (R_0 + j\delta))$ (D is for disk) and $A_j = D_j - D_{j-1}$ (A is for annulus). By repeating the one dimensional proof we can show that with high probability

(a) All the houses in D_0 stay reasonable until time T.

(b) The houses in A_j will be good at time jS and stay reasonable to time T.

(c) All the houses in $D_B \supset B_2(0, 3R_0)$ will be good at time T.

and the desired result follows from an application of Theorems 4.3 and 4.2 as before. □

We will now give the promised explanation of the conjecture for the case $\kappa = 2$. First consider the situation in $d = 1$ and for ease of exposition call the two states "yellow" and "blue". As our proof shows if we have a sufficiently large interval of sites in which two colors occur with approximately equal frequency then the distribution of colors in this region will quickly converge to a product measure with density $1/2$ and the region will expand, no matter what it encounters outside. For the region to expand we need $\theta = a|\mathcal{N}|$ with $a < 1/4$ for if $a > 1/4$ and all sites in $[r, 2r)$ are yellow then the random region cannot expand since the site at r will have about $r/2 < a(2r + 1)$ blue sites in its neighborhood. Applying the same reasoning to yellow sites in $x \in [br, r)$, who have about $(2 - b)r/2$ blue sites in their neighborhood, we see that if $(2 - b)/2 < 2a$, i.e. $b > 2 - 4a$ then the yellow sites in $[br, r)$ will not flip to blue but since $a < 1/2$ the blue sites will flip to yellow at rate one.

Similar reasoning applies to the system in $d > 1$ with $1/4 < a < 1/2$ and shows that a large enough ball of yellow sites will expand through a random region. The trouble with turning this into a proof is that we cannot guarantee that the blob will always find itself in competition with a random region. Indeed in a deterministic version of the threshold voter model in $d = 1$ (see Durrett and Steif (1993)) this naive picture is not correct since there are "blockades" that in some circumstances will stop the advance of blobs. However, we believe that this will not happen in random systems or in $d > 1$. In support of this claim, we note that Andjel, Mountford, and Liggett (1992) have shown that clustering occurs in $d = 1$ when $\mathcal{N} = \{-k, \ldots, k\}$ and $\theta = k$. The important special property of this example is that if an interval of 1's (or 0's) is long enough only the site on either end can flip.

7. Long Range Limits

In the last section, we saw that the cyclic color model and Greenberg Hastings models simplified considerably when the range of interaction was large. In this section we show that the contact process also simplifies in this way.

Example 7.1. The basic contact process. As usual the neighborhood is $\mathcal{N} = \{x : \|x\|_p \leq r\}$. We will write the contact process as a set valued process with the state = the set of sites occupied by particles and formulate the dynamics as follows:

(i) Each particle dies at rate 1, and gives birth at rate β.

(ii) A particle born at x is sent to a site y chosen at random from $x + \mathcal{N}$.

(iii) If y is vacant, it becomes occupied. If y is already occupied the birth has no effect.

If r is large and the contact process starts from a single occupied site then at least until the number of particles is a significant fraction of $|\mathcal{N}|$, the contact process will behave like a *branching random walk*, i.e., the process that obeys (i) and (ii) but allows any number of particles per site.

The total number of particles at time t in a branching random walk is a *branching process* – a Markov chain Z_t in which transitions from k to $k+1$ occur at rate $k\beta$ and transitions k to $k-1$ occur at rate k. Let $T_y = \inf\{t : Z_t = y\}$ and use P_x to denote the law of the branching process with $Z_0 = x$. Well known properties of the exponential distribution imply that

$$P_k(T_{k+1} < T_{k-1}) = \frac{\beta}{\beta+1} \quad \text{for } k > 0$$

so Z_t is a time change of an asymmetric random walk S_n that, when $k > 0$, makes transitions $k \to k+1$ with probability $\beta/(\beta+1)$ and $k \to k-1$ with probability $1/(\beta+1)$ and has 0 as an absorbing state, i.e., once $S_n = 0$ we will have $S_m = 0$ for all $m > n$. Using this observation and well known formulas for simple random walk it follows that

$$P_1(T_0 < \infty) = \begin{cases} 1 & \text{if } \beta \leq 1 \\ 1/\beta & \text{if } \beta > 1 \end{cases}$$

so the critical value of β for the survival of the branching process is 1.

The main result in this section is that as the range $r \to \infty$ the critical value for survival of the contact process converges to that of the branching process. Let $\tau^0 = \inf\{t : \xi_t^0 = \emptyset\}$ where ξ_t^0 denotes the contact process starting from a single particle at the origin, i.e., $\xi_0^0 = \{0\}$. Let $\beta_c = \inf\{\beta : P(\tau^0 = \infty) > 0\}$.

Theorem 7.1. As $r \to \infty$, $\beta_c \to 1$ and if $\beta > 1$

$$P(x \in \xi_\infty^1) \to \frac{\beta-1}{\beta}$$

Remark. Schonmann and Vares (1986) have shown that if we consider the basic contact process in d dimensions with $\mathcal{N} = \{x : \|x\|_1 = 1\}$ and we let $\beta = 2d\lambda$ then the conclusions of Theorem 7.1 and (7.18) below hold.

PROOF: To begin we note that we can construct the contact process from a branching random walk by suppressing births onto occupied sites. So we can define the contact process and the branching random walk on the same space so that the branching random walk always has more particles than the contaact process, and it follows that $\beta_c \geq 1$ for all r. To prove the rest of the result we note that taking $A = \mathbf{Z}^d$ and $B = \{0\}$ in the duality equation (5.3) gives

$$P(\xi_t^1 \cap \{0\} \neq \emptyset) = P(\xi_t^0 \cap \mathbf{Z}^d \neq \emptyset) = P(\tau^0 > t)$$

Letting $t \to \infty$ we have

(7.1)
$$P(0 \in \xi_\infty^1) = P(\tau^0 = \infty)$$

So to prove Theorem 8.1 it suffices to show that

(7.2) If $\beta > 1$ then $P(\tau^0 = \infty) \to (\beta - 1)/\beta$

for this implies that $\limsup_{r\to\infty} \beta_c \leq 1$. To prove (8.2) we scale space by dividing by r and consider the contact process on $\mathbf{Z}^d/r$ to facilitate taking the limit $r \to \infty$. Our approach will be to use the comparison theorem, so we let $I_k = k2Le_1 + [-L, L]^d$ and consider a modification of the contact process $\bar{\xi}_t$ in which births are not allowed outside $(-4L, 4L)^d$. The two key ingredients in the proof are

(7.3) Let $\delta > 0$. If we pick L large, set $T = L^2$, and pick K large then for $r \geq r_0$, $\bar{\xi}_T$ will have at least K particles in I_1 and in I_{-1} with probability at least $1 - \delta$ whenever $\bar{\xi}_0$ has at least K particles in I_0

(7.4) Consider the process starting from $\xi_0^0 = \{0\}$. If we pick S large then for $r \geq r_1 \geq r_0$, ξ_S^0 will have at least K particles in I_0 with with probability at least $((\beta - 1)/\beta) - \delta$

Once this is done (7.2) follows by using Theorem 4.3 to compare

$$X_n = \{m : |\xi_{S+nT}^0 \cap I_m| \geq K\}$$

with a one-dependent oriented percolation with density $\geq 1 - \delta$ and Theorem 4.1 to conclude that the cluster containing $(0, 0)$ in the percolation model will be infinite with probability at least $1 - 55\delta^{1/9}$. For these two facts imply that

$$P(\tau^0 = \infty) \geq \frac{\beta - 1}{\beta} - \delta - 55\delta^{1/9}$$

PROOF OF (7.3): The starting point is the observation that if we let $r \to \infty$ then the contact process on $\mathbf{Z}^d/r$ converges to a branching random walk η_t in which

(i) Each particle dies at rate 1, and gives birth at rate β.

(ii) A particle born at x is sent to a point y chosen at random from $\{y : \|y - x\|_p \leq 1\}$.

This should be intutively clear since if we start with one particle at 0, fix T and let $r \to \infty$ then the probability of a *collision* (birth onto an occupied site) by time T goes to 0 as $r \to \infty$, and the displacements of the individual particles converge to a uniform distribution on $\{y : \|y\|_p \leq 1\}$.

We will prove the convergence of the contact process on $\mathbf{Z}^d/r$ to the branching random walk later (see the "continuity argument" below). We have introduced this result now to motivate the first step of the proof, which is to prove the analogue of (7.3) for the branching random walk η_t, which is given in (7.12) below. Let η_t^x denote the branching random walk starting from $\eta_0^x = \{x\}$. To leave room for the limit $r \to \infty$ we consider $\bar{\eta}_t^x$ a modification of η_t^x in which particles that land outside $(-4L + 1, 4L - 1)^d$ are killed. Let $m(t, x, A) = E|\bar{\eta}_t^x \cap A|$ be the mean number of particles in A at time t for the modified branching random walk starting with a single particle at x. We claim that

$$(7.5) \qquad m(t, x, A) = e^{(\beta - 1)t} P(\bar{W}_t^x \in A)$$

where $\bar{W}_t^x$ is a random walk that starts at x, jumps at rate β, has jumps that are uniform on $\{y : \|y\|_p \leq 1\}$, and is killed when it lands outside $(-4L+1, 4L-1)^d$. To check this claim note that both sides of (7.5) satisfy the same differential equation: if $A \subset (-4L+1, 4L-1)^d$ then

$$\frac{dm(t, x, A)}{dt} = -m(t, x, A) + \int m(t, x, dy)\nu(A - y)$$

where $A - y = \{x - y : x \in A\}$ and ν is the uniform probability measure on $\{y : \|y\|_p \leq 1\}$.

Let $I_1' = 2Le_1 + [-L + 1, L - 1]^d$, i.e. I_1 shrunk by a little bit. Donsker's theorem implies that if $T = L^2$ and $x/L \to \theta \in [-1, 1]^d$

$$(7.6) \qquad P(\bar{W}_T^x \in I_1') \to \psi(\theta)$$

where $\psi(\theta) = P_\theta(B_t \in [-4, 4]^d$ for $t \leq 1$, $B_1 \in 2e_1 + [-1, 1]^d)$ and B_t is a constant multiple of d-dimensional Brownian motion. $\psi(\theta) > 0$ and is continuous, so a simple argument (suppose not and extract a convergent subsequence) shows

$$(7.7) \qquad \liminf_{L \to \infty} \left[\inf_{x \in [-L, L]^d} P(\bar{W}_T^x \in I_1') \right] \geq \inf_{\theta \in [-1, 1]^d} \psi(\theta) > 0.$$

It follows from (7.5)–(7.7) that we can pick L large enough so that

$$(7.8) \qquad \inf_{x \in [-L, L]^d} E|\bar{\eta}_T^x \cap I_1'| \geq 2.$$

Let $\bar{\eta}_t^A$ denote the modified branching random walk with $\bar{\eta}_0^A = A$. (7.8) implies

$$(7.9) \qquad E|\bar{\eta}_T^A \cap I_1'| \geq 2|A|$$

while an obvious comparison and a well known fact about branching processes (see Athreya and Ney (1972) for this and other facts about branching processes we will use) implies

$$(7.10) \qquad \mathrm{var}(\bar{\eta}_T^x \cap I_1') \leq E|\bar{\eta}_T^x \cap I_1'|^2 \leq E|\eta_T^x|^2 = C_T < \infty$$

Combining the last two conclusions and using Chebyshev's inequality it follows that if $A \subset [-L, L]^d$ has $|A| = K$ then

$$(7.11) \qquad P(|\bar{\eta}_T^A \cap I_1'| < K) \leq \frac{\mathrm{var}(|\bar{Z}_T^A \cap I_1'|)}{(2|A| - K)^2} \leq \frac{K \sup_x \mathrm{var}(|\bar{Z}_T^x \cap I_1'|)}{K^2} \leq \frac{C_T}{K}$$

From the last result it follows that

(7.12) If $\delta > 0$ and K is large then for any $A \subset [-L, L]^d$ with $|A| = K$.

$$P(|\bar{\eta}_T^A \cap I_1'| < K) \leq \delta/10$$

Continuity Argument. (7.12) shows that if $A \subset I_0$ has $|A| = K$ then with high probability $\bar{\eta}_T^A$ will have at least K particles in I_{-1} and in I_1. The next step is to prove the corresponding result for the contact process. To avoid some technicalities we will give the details only for the case in which $\mathcal{N} = \{z : \|z\|_\infty \leq r\}$ and then indicate the extension to $p < \infty$ in a remark after the proof.

Let $\bar{\xi}_t^A$ be a modification of the contact process with $\bar{\xi}_0^A = A$ in which births outside $(-4L, 4L)^d$ are not allowed. We begin by observing that the number of births up to time t in the contact process, V_t, is dominated by a branching process $\bar{V}_t$ in which births occur at rate β and deaths occur at rate 0. If $|A| = K$ then $E\bar{V}_t = Ke^{\beta t} < \infty$, so our comparison and Chebyshev's inequality imply

$$(7.13) \qquad P(V_T > r^{1/3}) \leq P(\bar{V}_T > r^{1/3}) \leq \frac{Ke^{\beta T}}{r^{1/3}} \to 0$$

since T is fixed and $r \to \infty$.

Let $G_1 = \{V_t \leq r^{1/3}\}$. Here G is for good event and the subscript indicates it is the first of several we will consider. When G_1 occurs, the probability of having a birth land on an occupied site (a "collision") is

$$(7.14) \qquad \leq r^{1/3} \frac{r^{1/3}}{(2r + 1)^d} \to 0$$

since there are at most $r^{1/3}$ births and even if all the particles are in $\{x : \|x\|_\infty \leq 1\}$ (on $\mathbf{Z}^d/r$) each birth has probability at most $r^{1/3}/(2r + 1)^d$ of landing on an occupied site. Let G_2 be the event that there are no collisions by time t.

To deal with the spatial location of particles, we will create a coupling of the displacements of the particles in the branching random walk to those of particles in the contact process. To couple the displacements we observe that if U is uniform on $\{y : \|y\|_\infty \leq 1\}$

and $\pi_r(x)$ is the closest point in $\mathbf{Z}/r^d$ to x (with some convention for breaking ties) then $U^r = \pi_r(U(1+1/2r))$ is uniform on $\mathcal{N}/r$.

Now if the U_i are the displacements of particles in the branching random walk, we will use the U_i^r for the displacements in the contact process. When our good events G_1 and G_2 occur, we have $G_3 =$ all of the points in the contact process ξ_s^A are within $r^{1/3}/r$ (in $\|\ \|_\infty$) of their counterparts in the branching process η_s^A. Passing to the truncated processes and noting that the branching particles are required to stay in $(-4L+1, 4L-1)^d$ for $0 \le s \le T$, while the contact process particles are required to stay in $(-4L, 4L)^d$, it follows that on G_3 we have $|\bar{\xi}_T^A \cap I_1| \ge |\bar{\eta}_T^A \cap I_1'|$ Combining the last observation with (7.12) gives (7.3). □

Remark. If $p < \infty$ then $U^r = \pi_r((1+1/2r)U)$ is not uniform on $\mathcal{N}/r$ but is within C/r of uniform in the total variation norm. In the last paragraph of the proof we then have $P(\|U_i - U_i^r\|_\infty > 1/r) \le C/r$, which since there are at most $r^{1/3}$ transitions on G_1, is good enough for the proof.

PROOF OF (7.4): By the continuity argument it is enough to show that we can pick S so that η_S^0 will have at least K particles in I_0 with probability at least $((\beta - 1)/\beta) - \delta/2$. However, this follows from

(7.16) If Ω_∞ is the event that the branching process does not die out, then for any $L > 0$ and $K < \infty$,

$$P(|\eta_t^0 \cap [-L, L]^d| < K, \Omega_\infty) \to 0$$

Indeed as Asmussen and Kaplan (1976) have shown (see Theorem 2 on p. 5)

(7.17) There is a constant $\sigma > 0$ so that

$$\sqrt{t}e^{-(\beta-1)t}|\eta_t^0 \cap [-L, L]^d \to W \cdot \frac{(2L+1)^d}{(2\pi\sigma^2)^{d/2}}$$

where $W = \lim_{t\to\infty} e^{-(\beta-1)t}|\eta_t^0| > 0$ a.s. on Ω_∞

This completes the proof of (7.4) and hence of (7.2). □

The argument just used on the long range contact process can also be applied to

Example 7.2. Succesional dynamics. We suppose that the set of states at each site are $0 =$ grass, $1 =$ a bush, $2 =$ a tree and formulate the dynamics as

$$c_0(x, \xi) = \delta_{\xi(x)}$$
$$c_1(x, \xi) = \lambda_1 n_1(x) \quad \text{if } c_i(x) = 0$$
$$c_2(x, \xi) = \lambda_2 n_2(x) \quad \text{if } c_i(x) \le 1$$

The title of this example and its formulation are based on the observation that if an area of land is cleared by a fire, then regowth will occur in three stages: first grass appears then small bushes and finally trees, with each species growing up through and replacing the previous one. With this in mind, we allow each type to give birth onto sites occupied by lower numbered types.

Theorem 7.2. Let $\beta_i = \lambda_i |\mathcal{N}|$. Suppose that $\beta_2 > \delta_2$ and

$$(\star) \qquad \beta_1 \cdot \frac{\delta_2}{\beta_2} > \delta_1 + \beta_2 \cdot \frac{\beta_2 - \delta_2}{\beta_2}$$

If r is large then there is a nontrivial translation invariant stationary distribution in which all three types have positive density.

SKETCH OF PROOF: The fact that the 2's do not feel the presence of the 1's implies that the set of sites occupied by 2's is a contact process. To construct a stationary distribution we start with the 2's in their upper invariant measure and we put 1's at all sites not occupied by 1's to get a process ξ_t^{12}. This is the analogue of starting an attractive system from all 1's and a result of Durrett and Moller imples that as $t \to \infty$, $\xi_t^{12} \Rightarrow \xi_\infty^{12}$ a translation invariant stationary distribution.

To prove that ξ_∞^{12} is nontrivial we will prove an analogue of (7.3). The first step is to prove the following result about the long range contact process (which is here considered as a subset of $\mathbf{Z}^d$)

(7.18) If $\beta > 1$ and $x \neq y$ then as $r \to \infty$

$$P(x, y \in \xi_\infty^1) \to \left(\frac{\beta - 1}{\beta} \right)^2$$

In words, the equilibrium distribution converges to a product measure as $r \to \infty$. Of course, the last conclusion only says that the sites are asymptotically pairwise independent, but the argument can easily be generalized to a finite number of x's.

PROOF: By duality (see the proof of (7.1))

$$P(x, y \in \xi_\infty^1) = P(\tau^x = \infty, \tau^y = \infty)$$

Our comparison of the contact process with a branching process at the beginning of the proof of Theorem 8.1 shows that $P(\tau^x = \infty) \leq (\beta - 1)/\beta$ for all r. If we pick K and L as in (7.3) and then pick S large as in (7.4) then for $r \geq r_1$ we have

$$\frac{\beta - 1}{\beta} + \delta \geq P(\tau^x > S)$$

$$\geq P(|\xi_S^x \cap [-L, L]| > K) \geq \frac{\beta - 1}{\beta} - \delta$$

Our choice of K and L and the comparison with oriented percolation shows that

$$P(|\xi_S^x \cap [-L, L]| > K, \tau^x < \infty) \leq 55\delta^{1/9}$$

Combining the last two estimates shows

$$|P(\tau^z = \infty) - P(\tau^z > S)| \le \delta + 55\delta^{1/9}$$

With this in hand the desired result follows easily since continuity argument shows that for any fixed S as $r \to \infty$

$$P(\tau^z > S, \tau^y > S) \to P(\eta_S^0 \ne \emptyset)^2 \qquad \qquad \square$$

Turning now to the heart of the proof we will again scale space by dividing by r and consider the contact process on $\mathbf{Z}^d/r$ to facilitate taking the limit. The approach we will take is a combination of that of Durrett and Swindle (1991) and Durrett and Schinazi (1993). We will concentrate on explaining the main ideas and refer the reader to those papers for the details. Pick $\rho > (\beta_2 - \delta_2)/\beta_2$ so that

$(\star\star)$ $$\beta_1(1 - \rho) > \delta_1 + \beta_2\rho$$

By dividing space into cubes of side δr then using (7.18) and the weak law one can prove that with high probability all sites in our space time box have at most $\rho|\mathcal{N}|$ neighbors in state 2. (Recall that the set of 2's at any time is distributed according to the upper invariant measure.) This means that a single 1 will have births that land on an occupied site at rate $\ge \beta_1(1 - \rho)$ while it dies at rate δ_1 and is smothered by a 2 at rate $\le \beta_2\rho$.

The inequality $(\star\star)$ implies that a single particle gives birth faster than it dies. If we start with a fixed number of 1's then in the limit $r \to \infty$ the 1's dominate a supercritical branching random walk. If this fixed number K is large and L and $T = L^2$ are chosen appropriately then for large r a truncated version of the process which is not allowed to give birth outside $(-4L, 4L)^d$ will with high probability have at least K particles in I_1 and in I_{-1} whenver the initial configuration has at least K particles in I_0.

The last result is an analogue of (8.3) but there is one problem. The event that $\xi_t(x) = 2$, which is the same as the survival of the dual contact process of 2's starting from (x, t), does not have a finite range of dependence. To avoid this problem we adopt the more liberal viewpoint that x is occupied by a 2 at time t if the dual process escapes from a certain space-time box. If the box is large enough the liberalization of the definition does not increase the density of 2's by enough to violate $(\star\star)$, we can verify the comparison assumptions of Theorem 4.3 and the desired result follows from Theorem 8.2.

8. Rapid Stirring Limits

The point of this section is that if we take a fixed interacting particle system, scale space by ϵ and "stir" the particles at rate ϵ^{-2} then as $\epsilon \to 0$ the particle system converges to the solution of a reaction diffusion equation. To be precise, we consider processes $\xi_t^\epsilon : \epsilon \mathbf{Z}^d \to \{0, 1, \ldots, \kappa - 1\}$ that evolve as follows

(i) there are *translation invariant finite range flip rates*

$$c_i(x, \xi) = h_i(\xi(x), \xi(x + \epsilon y_1), \ldots, \xi(x + \epsilon y_N))$$

(ii) *rapid stirring:* for each $x, y \in \epsilon \mathbf{Z}^d$ with $\|x - y\|_1 = \epsilon$ we exchange the values at x and y at rate ϵ^{-2}. That is, we change the configuration from ξ to $\xi^{x,y}$ where

$$\xi^{x,y}(y) = \xi(x) \qquad \xi^{x,y}(x) = \xi(y) \qquad \xi^{x,y}(z) = \xi(z) \quad z \neq x, y$$

The reader should note that in (i) changing ϵ scales the lattice but does not change the interaction between the sites. In (ii) we superimpose stirring in such a way that the individual values will be moving according to Brownian motions (run at rate 2) in the limit. The motivation for modifying the system in this way comes from the following *mean field limit theorem* of De Masi, Ferrari, and Lebowitz (1986). The derivation of such "hydrodynamic limits" has become a major enterprise (see e.g., Spohn (1991) or DeMasi and Presutti (1992)) but this particular result is rather easy to establish.

Theorem 8.1. Suppose $\xi_0^\epsilon(x)$ are independent and let $u_i^\epsilon(t, x) = P(\xi_t^\epsilon(x) = i)$. If $u_i^\epsilon(0, x) = g_i(x)$ is continuous then as $\epsilon \to 0$, $u_i^\epsilon(t, x) \to u_i(t, x)$ the bounded solution of

(8.1) $$\partial u_i / \partial t = \Delta u_i + f_i(u) \qquad u_i(0, x) = g_i(x)$$

where

(8.2) $$f_i(u) = \; < c_i(0, \xi) 1_{(\xi(0) \neq i)} >_u \; - \sum_{j \neq i} < c_j(0, \xi) 1_{(\xi(0) = i)} >_u$$

and $< \phi(\xi) >_u$ denotes the expected value of $\phi(\xi)$ under the product measure in which state j has density u_j, i.e., when $\xi(x)$ are i.i.d. with $P(\xi(x) = j) = u_j$.

Theorem 8.1 is easy to understand. The stirring mechanism (i.e., (ii)) has product measures as its stationary distributions. See Griffeath (1979), Section II.10. When ϵ is small, stirring operates at a fast rate and keeps the system close to a product measure. The rate of change of the densities can then be computed assuming adjacent sites are independent. To help explain the somewhat ugly formula in (8.2) we will now consider two concrete examples.

Example 8.1. The basic contact process. In this case $c_0(x, \xi) = 1$ and $c_1(x, \xi) = \lambda n_1(x, \xi)$ where $n_i(x, \xi) = |\{y \in \mathcal{N} : \xi(x + y) = i\}|$ is the number of neighbors in state i.

We claim that when $|\mathcal{N}| = N$ the equation in (9.1) becomes (we do not need an equation for $u_0 = 1 - u_1$)

$$\partial u_1/\partial t = \Delta u_1 - u_1 + N\lambda(1 - u_1)u_1$$

To see the second term on the right hand side the equation, we note that particles die at rate 1 independent of the state of neighbors. For the third, we note that if we assume all sites are independent then the probability x is vacant and $y \in x + \mathcal{N}$ is occupied is $(1 - u_1)u_1$. Each such pair produces a new particle at rate λ and there are N such pairs, so the total rate at which new particles are created (assuming that adjacent sites are independent) is $N\lambda(1 - u_1)u_1$.

The equation in the last example is just the mean field equation for the contact process that we have seen several times before. To see something new we look at

Example 8.2. The threshold one voter model. In this case

$$c_i(x, \xi) = 1 \quad \text{if } n_i(x, \xi) \geq 1$$

and if we assume $|\mathcal{N}| = N$ then the limiting equation is (again we do not need an equation for $u_0 = 1 - u_1$)

$$\partial u_1/\partial t = \Delta u_1 - u_1(1 - u_1^N) + (1 - u_1)(1 - (1 - u_1)^N)$$

To see this note that if all sites are independent then the probability x is occupied and at least one neighbor is vacant is $u_1(1 - u_1^N)$ and this is the rate at which 1's are destroyed. Intercahnging the roles of vacant and occupied in the last sentence gives the third term.

Having explained the formula in (8.2) we turn now to a result that extends Theorem 8.1 by showing that the particle system itself, not just its expected values are close to the p.d.e. To motivate the statement we note that the states of the sites in the model become independent in the limit $\epsilon \to 0$ and the number of sites per unit volume becomes large so it should not be surprising that in the limit $\xi_t^\epsilon(x)$ becomes deterministic.

Theorem 8.2. Let $\phi(x)$ be a smooth function with compact support. As $\epsilon \to 0$

$$\epsilon^d \sum_{y \in \epsilon \mathbf{Z}^d} \phi(y) 1_{(\xi_t^\epsilon(y)=i)} \to \int \phi(y) u_i(t, y)\, dy$$

in probability.

Although the indicator function of a bounded open set G is not continuous, this should be thought of as saying that

$$\epsilon^d \sum_{y \in \epsilon \mathbf{Z}^d \cap G} 1_{(\xi_t^\epsilon(y)=i)} \to \int_G u_i(t, y)\, dy$$

or more intuitively that the fraction of sites near y that are in state i converges to $u_i(t, y)$. The result for an open set G is also true, but is a little more difficult to prove precisely because 1_G is not continuous.

Theorem 8.2 provides a link between the particle system with fast stirring that we will exploit in the next lecture to prove the existence of stationary distributions for a predator prey model with fast stirring. Once Theorem 8.1 is established, the proof of Theorem 8.2 is easy: compute second moments and use Chebyshev's inequality. So we will concentrate on the proof of Theorem 8.1. The ideas behind the proof are simple: we will give an explicit construction of the process that allows us to define a dual processes by asking the question: "What is the state of x at time t?" and working backwards in time. The answer to this question can be determined by looking at the states of the sites in the "dual process" $I_\epsilon^{x,t}(s)$ at time $t - s$. The particles in $I_\epsilon^{x,t}(s)$ move according to stirring at a fast rate and give birth to new particles at rate

$$c^* = \sup_\xi \sum_i c_i(x, \xi)$$

We will show that for small ϵ the dual process is almost a branching random walk and converges to a branching Brownian motion as $\epsilon \to 0$. The proof of the last result leads easily to the conclusion that two dual processes $I_\epsilon^{x,t}(s)$ and $I_\epsilon^{y,t}(s)$ are asymptotically independent which gives the asymptotic independence of the sites in the particle systems. The convergence of the dual process to branching Brownian motion leads in a straightforward way to the convergence of the $u_i^\epsilon(t, x)$ to limits $u_i(t, x)$ and the asymptotic independence of adjacent sites implies that the $u_i(t, x)$ satisfy the limiting equations.

a. The dual process. The first step in the proof is to construct the process from a number of Poisson processes, all of which are assumed to be independent. The construction is similar in spirit to the one in Section 3 but it is convenient to do the details in a slightly different way. For each $x \in \epsilon \mathbf{Z}^d$, let $\{T_n^x, n \geq 1\}$, be a Poisson process with rate c^* and let $\{U_n^x, n \geq 1\}$ be a sequence of independent random variables that are uniform on $(0, 1)$. At time T_n^x we compute the flip rates $r_i = c_i(x, \xi(T_n^x))$ and use U_n^x to determine what (if any) flip should occur at x at time T_n^x. To be precise we let $p_i = \sum_{j \leq i} r_j / c^*$ for $i = 0, \dots, \kappa - 1$ with $p_{-1} = 0$ and flip to i if $U_n^x \in (p_{i-1}, p_i)$. If $U_n^x \in (p_{\kappa-1}, 1)$ no flip occurs. To move the particles around, we let $\{S_n^{x,y}, n \geq 1\}$ be Poisson processes with rate ϵ^{-2} when $x, y \in \epsilon \mathbf{Z}^d$ with $\|x - y\|_1 = \epsilon$, and we declare that at time $S_n^{x,y}$ the values at x and y are exchanged.

The dual process $I_\epsilon^{x,t}(s)$ is naturally defined only for $0 \leq s \leq t$ but for a number of reasons, it is convenient to assume that the Poisson processes and uniform random variables in the construction are defined for negative times and define $I_\epsilon^{x,t}(s)$ for all $s \geq 0$. Let $\mathcal{N} = \{\epsilon y_1, \dots, \epsilon y_N\}$ be the set of neighbors of 0. The dual process makes transitions as follows:

If $y \in I_\epsilon^{x,t}(s)$ and $T_n^y = t - s$ then we add all the points of $y + \mathcal{N}$ to $I_\epsilon^{x,t}(s)$.

If $y \in I_\epsilon^{x,t}(s)$ and $S_n^{y,z} = t - s$ then we move the particle at y to z.

For a picture of (a rather unlikely sample path for) the dual when $d = 1$ and $\mathcal{N} = \{-1, 0, 1\}$ see Figure 8.1

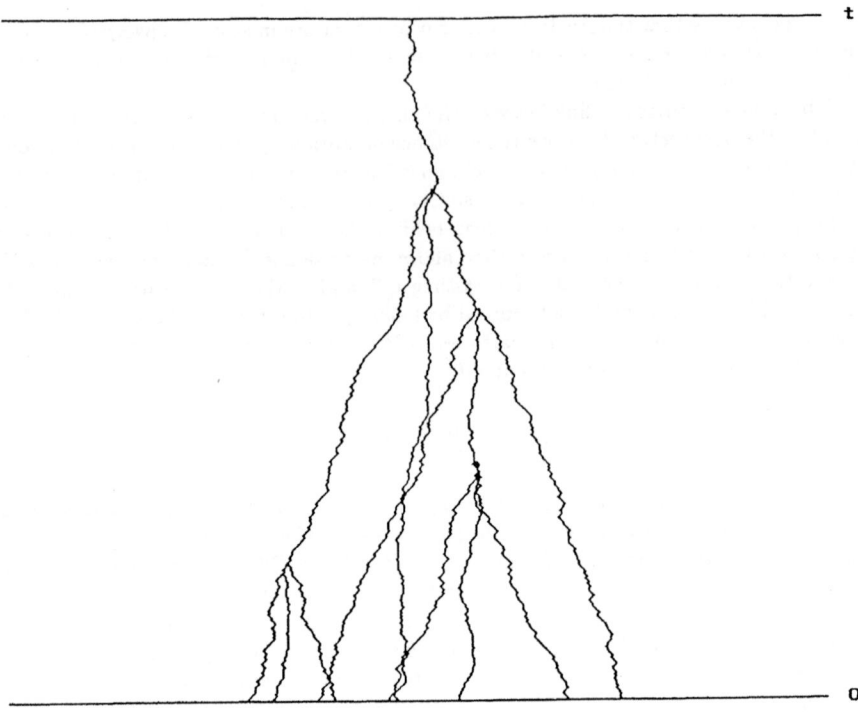

Figure 8.1

It is easy to see that we can compute the state of x at time t by knowing the states of the y in $I_\epsilon^{x,t}(s)$ at time $t - s$. We start with the values in $I_\epsilon^{x,t}(s)$ at time $t - s$ and work up to time t. At S arrivals we perform the indicated stirrings. When an arrival T_n^y occurs at a point of the dual, we look at the value of the process on $y + \mathcal{N}$, compute the flip rates r_i, and use U_n^x to determine what (if any) flip should occur.

To prepare for the proof of the convergence of $u_i^\epsilon(t, x)$ we will now give a more detailed description of $I_\epsilon^{x,t}(s)$. Let $X_\epsilon^0(0) = x$, let R_ϵ^1 be the smallest value of s so that we have a T arrival at $X_\epsilon^0(s)$ at time $t - s$, and set $X_\epsilon^i(s) = \epsilon y_i + X_\epsilon^0(s)$ for $1 \le i \le N$. Finally, we set $\mu_\epsilon^1 = 0$ to indicate that 0 is the mother of the N particles created at time R_ϵ^1. Passing now to the inductive step of the definition, suppose that we have defined the process up to time R_ϵ^m with $m \ge 1$. The $mN + 1$ existing particles move as dictated by stirring until R_ϵ^{m+1}, the first time $s > R_\epsilon^m$ that a T arrival occurs at the location of one of our moving particles $X_\epsilon^k(s)$ and then we set $X_\epsilon^{mN+i}(s) = \epsilon y_i + X_\epsilon^k(s)$ for $1 \le i \le N$, and $\mu_\epsilon^{m+1} = k$. The new particles may be created at the locations of existing particles. If so we say that a *collision* occurs and call the new particle *fictitious*. We will prove later that the probability of a collision tends to 0 as $\epsilon \to 0$, but for proving the convergence of $u_i^\epsilon(t, x)$, it is convenient to allow the fictitious particles to move and give birth like other particles, so for each $m \geq 1$ we define an independent copy of the graphical representation which we use for the births and movement of the mth particle if it is fictitious. By definition all the offspring of fictitious particles are also fictitious.

b. The dual process is almost a branching random walk. The point of introducing fictitious particles is that $\mathcal{K}_t = mN + 1$ for $t \in [R_m^\epsilon, R_{m+1}^\epsilon)$ defines a branching process in which each particle gives birth to N additional particles at rate c^*. Our next goal is to show that if ϵ is small then $I_\epsilon^{x,t}(s)$ is almost a branching random walk in which particles jump to a randomly chosen neighbor at rate $2d\epsilon^{-2}$ and give birth as above. To do this we will couple X_ϵ^k to independent random walks Y_ϵ^k that start at the same location at time $\beta_k =$ the birth time of X_ϵ^k, and jump to a randomly chosen neighbor at rate $2d\epsilon^{-2}$.

We say X_ϵ^k is *crowded* at time s if for some $j \neq k$ $\|X_\epsilon^j(s) - X_\epsilon^k(s)\|_1 \leq \epsilon$. When X_ϵ^k is not crowded, we define the displacements of Y_ϵ^k to be equal to those of X_ϵ^k. When X_ϵ^k is crowded we use independent Poisson processes to determine the jumps of Y_ϵ^k. To estimate the difference between X_ϵ^k and Y_ϵ^k, we need to estimate the amount of time X_ϵ^k is crowded. Let $j \neq k$, $V_s^\epsilon = X_\epsilon^k(s) - X_\epsilon^j(s)$ and W_s^ϵ be a random walk that jumps to a randomly chosen neighbor at rate $4d\epsilon^{-2}$. (Notice that V_s^ϵ is the difference of two random walks and hence jumps at rate $4d\epsilon^2$. The transition probabilities of V_s^ϵ differ slightly from those of W_s^ϵ when $\|x\|_1 = \epsilon$. Here y denotes any point $\neq -x$ with $\|y\|_1 = \epsilon$.

jumps from x to	rate in V	rate in W
$-x$	ϵ^{-2}	0
0	0	$2\epsilon^{-2}$
$x + y$	$2\epsilon^{-2}$	$2\epsilon^{-2}$

From the last table it should be clear that $\|W_s^\epsilon\|_1$ is stochastically smaller than $\|V_s^\epsilon\|_1$, i.e., the two random variables can be constructed on the same space so that $\|W_s^\epsilon\|_1 \leq \|V_s^\epsilon\|_1$ for all s. To check this note that all the transition of V and W can be coupled except those in the first two lines of the table, but there $\|W\|_1$ jumps from 1 to 0 at rate ϵ^{-2} while $\|V\|_1$ jumps from 1 to 1 at rate $\epsilon^2/2$.

From the last comparison of $\|V\|_1$ and $\|W\|_1$ it follows that for any integer $M \geq 1$, $v_t^{M\epsilon} = |\{s \leq t : \|V_s^\epsilon\|_1 \leq M\epsilon\}|$ is stochastically smaller than $w_t^{M\epsilon} = |\{s \leq t : \|W_s^\epsilon\|_1 \leq M\epsilon\}|$. Well known asymptotic results for random walks imply that when $t\epsilon^{-2} \geq 2$

$$(8.3) \qquad Ew_t^{M\epsilon} \leq \begin{cases} CM^d\epsilon^2 & d \geq 3 \\ CM^2\epsilon^2 \log(t\epsilon^{-2}) & d = 2 \\ CM\epsilon t^{1/2} & d = 1 \end{cases}$$

To see this note that $w_t^{M\epsilon}$ has the same distribution as $\epsilon^2 w_{t\epsilon^{-2}}^M$ and the last line is equal to $CM\epsilon^2(t\epsilon^{-2})^{1/2}$.

Let $\chi_\epsilon^k(t)$ be the amount of time X_ϵ^k is crowded in $[0, t]$. It is easy to see that

$$(8.4) \qquad E(\chi_\epsilon^k(t)|\mathcal{K}_t = K) \leq KEw_t^\epsilon$$

$$(8.5) \qquad E\mathcal{K}_t = \exp(\nu t) \text{ where } \nu = c^*N$$

$$(8.6) \qquad E(\chi_\epsilon^k(t)) \leq \exp(\nu t)Ew_t^\epsilon$$

To estimate the difference between $X_\epsilon^k(s)$ and $Y_\epsilon^k(s)$ we observe that if $\chi_\epsilon^k(t) = \tau$ then the number of "independent jumps" in the ith coordinate of Y_ϵ^k that occur in $[0, t]$ has a Poisson distribution with mean $\epsilon^{-2}\tau$. Let $\Delta_Y^i(s)$ be the net effect of the independent jumps on coordinate i up to time s. Recalling that changes in the ith coordinate of Y_ϵ^k have mean 0 and variance ϵ^2, it follows that $E\Delta_Y^i(s) = 0$ and

$$(8.7) \qquad E(\Delta_Y^i(s)^2) = E\chi_\epsilon^k(s)$$

Since $\Delta_Y^i(s)$ is a martingale, Kolmogorov's inequality implies

$$(8.8) \qquad E\left(\max_{0 \le s \le t} \Delta_Y^i(s)^2\right) \le 4E(\Delta_Y^i(t)^2)$$

Using Markov's inequality (i.e., if $X \ge 0$ then $P(X > x) \le EX^r/x^r$) then (8.8), (8.7), (8.6), and (8.3) (noting that the worst case is $d = 1$) gives

$$(8.9) \qquad P\left(\max_{0 \le s \le t} |\Delta_Y^i(s)| \ge \epsilon^{.3}\right) \le \epsilon^{-.6} E\left(\max_{0 \le s \le t} \Delta_Y^i(s)^2\right) \le C\epsilon^{.4}t^{1/2}\exp(\nu t)$$

Here and in what follows C dentores a constant whose value is unimportant and that will change from line to line. The arguments leading to the last inequality also apply to $\Delta_X^i(t)$, the net effect of jumps in $[0, t]$ while X_ϵ^k is crowded, so

$$(8.10) \qquad P\left(\max_{0 \le s \le t} \|X_\epsilon^k(s) - Y_\epsilon^k(s)\|_\infty \ge 2\epsilon^{.3}\right) \le C\epsilon^{.4}t^{1/2}\exp(\nu t)$$

The estimate in (8.10) shows that the X_ϵ^k are close to independent random walks. To see that with high probability no collisions occur, we pick M large enough so that $\|x\|_1 \le M$ for all $x \in \mathcal{N}$ and repeat the derivation of (8.6) with ϵ replaced by $M\epsilon$ to conclude that the expected number of births from X_ϵ^k while there is some other X_ϵ^j in $X_\epsilon^k + \mathcal{N}$ is smaller than

$$(8.11) \qquad C\epsilon t^{1/2}\exp(\nu t)$$

(8.5) and Markov's inequality imply that

$$(8.12) \qquad P(\mathcal{K}_t > K) \le K^{-1}\exp(\nu t)$$

When $\mathcal{K}_t \le K$, (8.11) implies that the expected number of collisions is smaller than

$$(8.13) \qquad KC\epsilon t^{1/2}\exp(\nu t)$$

Combining the last two results and setting $K = \epsilon^{-.5}$ shows that the probability of a collision is smaller than

$$(8.14) \qquad C\epsilon^{.5}(1 + t)^{1/2}\exp(\nu t)$$

Having shown that collisions are unlikely we no longer have to worry about the labels μ_m^ϵ that tell us the mother of the N particles created at time R_m^ϵ since this will be clear from the evolution of the dual. A more significant consequence of the results in this subsection is that dual processes for different sites are asymptotically independent. To argue this, we say the two duals *collide* if a particle in one dual gives birth when crowded by a particle in the other one. The arguments leading to (8.14) show that with high probability two duals do not collide, and (8.10) implies that the movements of all the particles can be coupled to independent random walks.

c. **Convergence of $u_i^\epsilon(t,x)$.** The next step is to show that as $\epsilon \to 0$ the branching random walk Y converges to a branching Brownian motion Z. To do this we use Skorokhod's trick to embed the ith component of the kth walk, $Y_s^{k,i}$ in a a Brownian motion $Z_s^{k,i}$. Using some standard estimates (see Durrett and Neuhauser for details) it follows that

$$(8.15) \qquad P\left(\max_{0 \le s \le t} \|Y_\epsilon^k(s) - Z^k(s)\|_\infty > 4\epsilon^{.3} \text{ for some } k \le K\right) \le KC\epsilon^{.32}(1+t)$$

To compute the state of x at time t, we need not only the dual process $I_\epsilon^{x,t}(s)$, $s \le t$ but also the labels μ_n^ϵ and the uniform random variables U_n^x. However, the uniform random variables are independent of the dual process and, as we pointed out in a remark after (8.14), the μ_n^ϵ are only needed when a collision occurs.

As we will now explain, the results in the last paragraph make it easy to show that $u_i^\epsilon(t,x) \to u_a(t,x)$ as $\epsilon \to 0$. Here and in what follows we will use a and b to denote possible states of the sites to ease the burden on the middle of the alphabet. The first step is to describe $u_a(t,x)$. Let Z_s be a branching Brownian motion starting with a single particle at x and let $\mathcal{K}_t$ be the number of particles at time t. For $0 \le k < \mathcal{K}_t$, we let $\zeta_0(k)$ be independent and $= a$ with probability $\phi_a(Z_t^k)$. Once the ζ_0 are defined, we work up the space time set $\{Z_{t-s}^k\} \times \{s\}$. The values of $\zeta_s(k)$, the state of Z_{t-s}^k at time s, stay constant as long as only stirring occurs. When $N+1$ branches Z_{t-s}^i, $Z_{t-s}^{kN+1}, \dots Z_{t-s}^{(k+1)N}$ come together (corresponding to a birth in the dual), we compute the flip rate at Z_{t-s}^i assuming it is in state $\zeta_s(i)$ and its neighbors are in states $\zeta_s(kN+j), 1 \le j \le N$. We generate an independent random variable uniform on $(0,1)$ to determine what (if any) flip should occur at Z_{t-s}^i. After we decide if we should change $\zeta_s(i)$, we can ignore $\zeta_s(kN+j)$ for $1 \le j \le N$. When we reach time t we will only be looking at the value at $\zeta_t(0)$. We call this value, the *result of the computation* and let $u_a(t,x) = P(\zeta_t(0) = a)$.

The description in the last paragraph is much like the one given earlier for the dual with one exception: the uniform random variables come from an auxiliary i.i.d. sequence instead of being read off the graphical representation. When there are no collisions in the dual, then the family structure of the influence set and the branching Brownian motion are the same. In this case if the inputs $\zeta_0(k)$ and the uniform random variables used are the same, the two computations have the same result. We have supposed that the initial functions $\phi_b(x)$ are continuous so (2.19) implies that as $\epsilon \to 0$,

$$\max_k |\phi_b(X_\epsilon^k(t)) - \phi_b(Z^k(t))| \to 0$$

where the maximum is taken over particles alive at time t. The last observation implies that we can with high probability arrange for all the inputs to be the same and it follows that $u_a^\epsilon(t, x) \to u_a(t, x)$. The last proof extends trivially to show that if $x_\epsilon \to x$ then $u_a^\epsilon(t, x_\epsilon) \to u_a(t, x)$. At the end of subsection b, we observed that the influence sets from different points are asymptotically independent. Combining that observation with the proofs in this subsection implies that if $x_\epsilon \to x$ then

$$(8.16) \qquad P(\xi_t^\epsilon(x_\epsilon + \epsilon y_j) = c_j, 0 \le j \le N) \to \prod_{j=0}^{N} u_{c_j}(t, x)$$

We are interested in statements that allow $x_\epsilon \to x$ since this form of the conclusion implies that convergence occurs uniformly on compact sets.

d. The limit satisfies the p.d.e. The first step is to write the limiting equation in integral form.

(8.17) **Lemma.** Suppose $f_a, 0 \le a < \kappa$ are continuous and $g_a, 0 \le a < \kappa$ are bounded and continuous. The following statements are equivalent:
(i) The functions $u_a(t, x)$ are a classical solution of

$$\frac{\partial u_a}{\partial t} = \Delta u_a - f_a(u) \qquad u_a(0, x) = g_a(x)$$

i.e., the indicated derivatives exist and are continuous.
(ii) The functions $u_a(t, x)$ are bounded and satisfy

$$u_a(t, x) = \int p_t(x, y) g_a(y) dy + \int_0^t ds \int p_s(x, y) f_a(u(t - s, y)) dy$$

where $p_t(x, y)$ is the transition probability for Brownian motion run at rate 2.

Proof: (i) implies that $Z_s^a \equiv u_a(t - s, B_s) - \int_0^s f_a(u(t - r, B_r)) dr$ is a bounded martingale, so $Z_0^a = EZ_t^a$ and (ii) follows from Fubini's theorem. To prove the converse, we begin by observing that if (ii) holds then $u_a(t, x)$ has the necessary derivatives and Z_s^a is a martingale, so (i) follows from Itô's formula. □

To get (ii) we will use the integration by parts formula. Let S_t^ϵ be the semigroup for the stirred particle system and T_t^ϵ be the semigroup for pure stirring. The integration by parts formula implies that for nice functions ψ we have

$$(8.18) \qquad S_t^\epsilon \psi(\xi) = T_t^\epsilon \psi(\xi) + \int_0^t ds\, S_{t-s}^\epsilon L T_s^\epsilon \psi(\xi)$$

where L is the generator for the particle system with no stirring. We use (8.18) with $\psi_{x,a}(\xi) = 1$ if $\xi(x) = a$ and 0 otherwise. Now for this choice of ψ

$$(8.19) \qquad T_s^\epsilon \psi_{x,a}(\xi) = \sum_y p_s^\epsilon(x, y) \psi_{y,a}(\xi)$$

where $p_s^\epsilon(x, y)$ is the transition probability of a random walk that jumps from y to z at rate $\epsilon^{-2}/2$ if $\|y - z\|_1 = \epsilon$. Now if $c_b(y, \xi) = h_b(\xi(y + \epsilon y_0), \ldots, \xi(y + \epsilon y_N))$ then

$$(8.20) \qquad L\psi_{y,a} = -\sum_b h_{b_0}(a, b_1, \ldots, b_N)\psi_{y,a} \prod_{j=1}^N \psi_{y+\epsilon y_j, b_j}$$

$$+ \sum_b h_a(b_0, b_1, \ldots, b_N)\psi_{y,b_0} \prod_{j=1}^N \psi_{y+\epsilon y_j, b_j}$$

where the sums are over $b_0, \ldots, b_N \in \{0, 1, \ldots, \kappa - 1\}$. Substituting (8.19) and (8.20) into (8.18) gives

$$(8.21) \quad P(\xi_t^\epsilon(x) = a) = \sum_y p_t^\epsilon(x, y)g_a(y)$$

$$+ \int_0^t ds \sum_y p_s^\epsilon(x, y) E\left\{ -\sum_b h_{b_0}(a, b_1, \ldots, b_N)\psi_{y,a}(\xi_{t-s}^\epsilon) \prod_{j=1}^N \psi_{y+\epsilon y_j, b_j}(\xi_{t-s}^\epsilon) \right.$$

$$\left. + \sum_b h_a(b_0, b_1, \ldots, b_N)\psi_{y,b_0}(\xi_{t-s}^\epsilon) \prod_{j=1}^N \psi_{y+\epsilon y_j, b_j}(\xi_{t-s}^\epsilon) \right\}$$

The local central limit theorem implies

$$(8.22) \qquad \sum_y |\epsilon^d p_s(x, y) - p_s^\epsilon(x, y)| \to 0$$

as $\epsilon \to 0$. As we observed at the end of subsection c,

$$E\psi_{y,c_0}(\xi_{t-s}^\epsilon) \prod_{j=1}^N \psi_{y+\epsilon y_j, c_j}(\xi_{t-s}^\epsilon) \to \prod_{j=0}^N u_{c_j}(t - s, y)$$

and this convergence occurs uniformly on compact sets. Using (8.21), (8.22), and the dominated convergence theorem, gives

$$(8.23) \quad u_a(t, x) = \int p_t(x, y)g_a(y)\, dy$$

$$+ \int_0^t ds \int dy\, p_s(x, y)\left\{ -\sum_b h_{b_0}(a, b_1, \ldots, b_N)u_a(t - s, y) \prod_{j=1}^N u_{b_j}(t - s, y) \right.$$

$$\left. + \sum_b h_a(b_0, b_1, \ldots, b_N)u_{b_0}(t - s, y) \prod_{j=1}^N u_{b_j}(t - s, y) \right\}$$

The term in braces is

$$(9.24) \qquad -\sum_{b \neq a} <c_b(0, \xi)1_{\{\xi(0)=a\}}>_{u(t-s,y)} + <c_a(0, \xi)>_{u(t-s,y)} = f_a(u(t - s, y))$$

Combining this with (8.17) gives the conclusion of Theorem 8.1.

9. Predator Prey Systems

In this section we will show that if you "know enough" about the limiting p.d.e. in Theorem 8.1 then you can prove results about the existence of stationary distributions for the system with fast stirring. For our approach, what you need to know about the p.d.e. is the following:

($\star$) There are constants $A_i < a_i < b_i < B_i$, L, and T so that if $u_i(0, x) \in (A_i, B_i)$ when $x \in [-L, L]^d$ then $u_i(x, T) \in (a_i, b_i)$ when $x \in [-3L, 3L]^d$.

Theorem 9.1. If ($\star$) holds then there is a nontrivial translation invariant stationary distribution for the process with fast stirring.

As the reader can probably guess, ($\star$) and Theorem 9.2 combine to produce a block event that turns one "pile of particles" into two and has high probability when ϵ is small and then the result follows from our comparison theorem. The details are somewhat technical so we refer the reader to Section 3 of Durrett and Neuhauser (1993) and turn to the problem of checking that ($\star$) holds in one particular example. For other applications of this technique see Durrett and Neuhauser (1993) or Durrett and Swindle (1993).

Example 9.1. Predator Prey Systems. The state at time t is $\xi_t^\epsilon : \epsilon \mathbf{Z}^d \to \{0, 1, 2\}$. We think of 0 as vacant, 1 and 2 as occupied by a fish and shark respectively. As usual, $n_i(x, \xi)$ is the number of neighbors of x (i.e., y with $\|y - x\|_1 = \epsilon$) that are in state i. The system changes states at the following rates:

$$\begin{aligned} c_1(x, \xi) &= \beta_1 n_1(x, \xi)/2d &&\text{if } \xi(x) = 0 \\ c_0(x, \xi) &= \delta_1 &&\text{if } \xi(x) = 1 \\ c_2(x, \xi) &= \beta_2 n_2(x, \xi)/2d &&\text{if } \xi(x) = 1 \\ c_0(x, \xi) &= \delta_2 + \gamma n_2(x, \xi)/2d &&\text{if } \xi(x) = 2 \end{aligned}$$

The first two rates say that fish repopulate vacant sites at a rate proportional to the number of fish at neighboring sites and die at rate δ_1. That is, in the absence of sharks, the fish are a contact process. The third rate says that sharks reproduce when they eat fish. This transition is a little strange from a biological point of view, but it has the desirable property that sharks will die out when the density of fish is too small. The last rate says that sharks die at rate δ_2 when they are isolated and the rate increases linearly with crowding. Finally, the sharks and fish swim around: for each pair of neighbors x and y stirring occurs at rate ϵ^{-2}, i.e., the values at x and y are exchanged. Applying Theorem 8.1 gives

Theorem 9.2. Suppose that $\xi_0^\epsilon(x)$, $x \in \epsilon \mathbf{Z}^d$ are independent and $u_i^\epsilon(t, x) = P(\xi_t^\epsilon(x) = i)$ for $i = 1, 2$. If $u_i^\epsilon(0, x) = \phi_i(x)$, which is continuous, then as $\epsilon \to 0$, $u_i^\epsilon(t, x) \to u_i(t, x)$ the bounded solution of

(9.1)
$$\begin{aligned} \frac{\partial u_1}{\partial t} &= \Delta u_1 + \beta_1 u_1(1 - u_1 - u_2) - \beta_2 u_1 u_2 - \delta_1 u_1 \\ \frac{\partial u_2}{\partial t} &= \Delta u_2 + \beta_2 u_1 u_2 - u_2(\delta_2 + \gamma u_2) \end{aligned}$$

with $u_i(0, x) = \phi_i(x)$.

As in the two examples in Section 8, the reaction terms are computed by assuming that adjacent sites are independent. To get $\beta_1 u_1(1 - u_1 - u_2)$ for example we note that if x is vacant and neighbor y is occupied by a fish, an event of probability $(1 - u_1 - u_2)u_1$ when sites are independent, births from y to x occur at rate $\beta_1/2d$ and there are $2d$ such pairs.

When the initial functions $\phi_i(x)$ are constant, $u_i(t, x) = v_i(t)$ and the v_i's satisfy

(9.2)
$$\frac{dv_1}{dt} = v_1((\beta_1 - \delta_1) - \beta_1 v_1 - (\beta_1 + \beta_2)v_2)$$
$$\frac{dv_2}{dt} = v_2(-\delta_2 + \beta_2 v_1 - \gamma v_2)$$

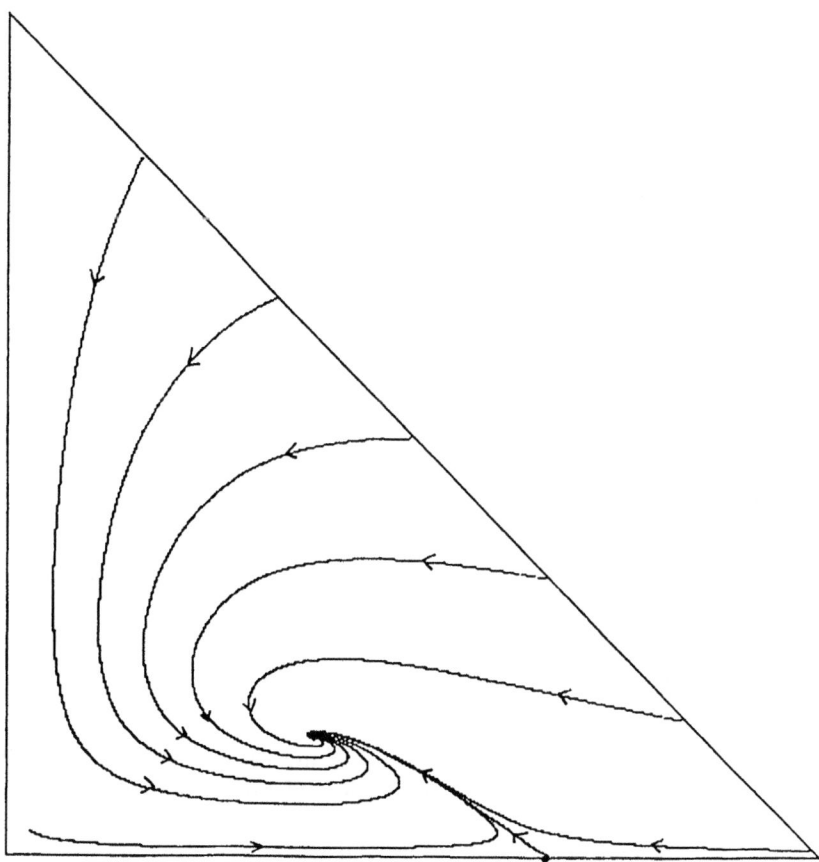

Figure 9.1

Here we have re-arranged the right hand side to show that the system is the standard predator–prey equations for species with limited growth. (See e.g., page 263 of Hirsch and Smale (1974).) Before we plunge into the details of analyzing (9.2), the reader should look at Figure 9.1, which gives some solutions of (9.2) with

$$\beta_1 = 3, \delta_1 = 1, \quad \beta_2 = 3, \delta_2 = 1, \quad \gamma = 1$$

In this case, as we will prove later, there is a fixed point at $(8/21, 3/21)$ that is globally attracting.

The first step in understanding (9.1) is to look at (9.2) and ask: "What are the fixed points, i.e., solutions of the form $v_i(t) = \rho_i$?" It is easy to solve for the ρ_i. There is always the trivial solution $\rho_1 = \rho_2 = 0$. In the absence of sharks the fish are a contact process. So if $\beta_1 > \delta_1$ there is a solution $\rho_1 = (\beta_1 - \delta_1)/\beta_1$, $\rho_2 = 0$. If we impose the stronger condition

(9.3) $$(\beta_1 - \delta_1)/\beta_1 > \delta_2/\beta_2$$

there is exactly one solution with $\rho_2 > 0$:

$$\rho_1 = \frac{(\beta_1 - \delta_1)\gamma + \delta_2(\beta_1 + \beta_2)}{\beta_1\gamma + \beta_2(\beta_1 + \beta_2)} \qquad \rho_2 = \frac{(\beta_1 - \delta_1)\beta_2 - \delta_2\beta_1}{\beta_1\gamma + (\beta_1 + \beta_2)\beta_2}$$

The condition $\beta_1 > \delta_1$ is an obvious necessary condition for the fish to survive in the absence of sharks. The condition (9.3) is not so intuitive but turns out to be sufficient for the existence of nontrivial stationary distributions for small ϵ.

Theorem 9.3. *Suppose that $(\beta_1 - \delta_1)/\beta_1 > \delta_2/\beta_2$ holds. If ϵ is small there is a nontrivial translation invariant stationary distribution in which the density of sites of type i is close to ρ_i.*

In view of Theorem 9.1 it suffices to prove $(\star)$, which is a consequence of the following convergence theorem.

Theorem 9.4. *Suppose that $(\beta_1 - \delta_1)/\beta_1 > \delta_2/\beta_2$ holds and the u_i solve (9.1) for continuous nonnegative $\phi_i(x)$ with $\phi_1(x) + \phi_2(x) \leq 1$ and $\phi_i(x_i) > 0$ for some x_i. Then there is a $\sigma > 0$ so that as $t \to \infty$,*

$$\sup_{\|x\| \leq \sigma t} |u_i(t, x) - \rho_i| \to 0.$$

PROOF: The proof is based on a simple idea due to Redheffer, Redlinger, and Walter (1988): the existence of a convex strict Lyapunov function for the dynamical system (10.2) plus two technical conditions we will identify in the proof, give a convergence theorem for the reaction diffusion equation. In this case the desired function is

$$H(v_1, v_2) = \beta_2(v_1 - \rho_1 \log v_1) + (\beta_1 + \beta_2)(v_2 - \rho_2 \log v_2)$$

Being the sum of four convex functions, H is clearly convex. The next step is to check that it is a strict Lyapunov function: if (v_1, v_2) is a solution of the dynamical system that does not start at the fixed point then $\partial H(v_1, v_2)/\partial t < 0$. Differentiating gives

$$\frac{\partial H}{\partial v_1} = \beta_2 \left(1 - \frac{\rho_1}{v_1} \right) \qquad \frac{\partial H}{\partial v_2} = (\beta_1 + \beta_2)\left(1 - \frac{\rho_2}{v_2} \right)$$

So using the chain rule and (9.2)

$$H_t \equiv \frac{\partial H(v_1, v_2)}{\partial t} = \beta_2(v_1 - \rho_1)\{(\beta_1 - \delta_1) - \beta_1 v_1 - (\beta_1 + \beta_2)v_2\}$$
$$+ (\beta_1 + \beta_2)(v_2 - \rho_2)\{-\delta_2 + \beta_2 v_1 - \gamma v_2\}$$

Using the next two identities to subtract 0 from each term in braces

$$0 = (\beta_1 - \delta_1) - \beta_1 \rho_1 - (\beta_1 + \beta_2)\rho_2$$
$$0 = -\delta_2 + \beta_2 \rho_1 - \gamma \rho_2$$

gives

(9.4)
$$H_t = \beta_2(v_1 - \rho_1)\{-\beta_1(v_1 - \rho_1) - (\beta_1 + \beta_2)(v_2 - \rho_2)\}$$
$$+ (\beta_1 + \beta_2)(v_2 - \rho_2)\{\beta_2(v_1 - \rho_1) - \gamma(v_2 - \rho_2)\}$$
$$= -\beta_1\beta_2(v_1 - \rho_1)^2 - \gamma(\beta_1 + \beta_2)(v_2 - \rho_2)^2 \le 0$$

with strict inequality for $(v_1, v_2) \ne (\rho_1, \rho_2)$. The importance of the last conclusion is that $H(v_1(t), v_2(t))$ is strictly decreasing in t and hence all trajectories that begin in $(0, \infty)^2$ must end at the minimum of H, (ρ_1, ρ_2). For later purposes we would like to note that the level curves $H_t = -r$ are concentric ellipses.

The above computations that show H is a Lyapunov function obviously depend on the special form of (9.2). To prepare for other applications at the end of this section, we would like the reader to check that in what follows only equations (9.6) and (9.10) depend on the special form of H.

Since composing the Lyapunov function with solutions of the dynamical system shows that they converge to the fixed point, it is natural to look at $h(t, x) = H(u_1(t, x), u_2(t, x)) - H(\rho_1, \rho_2) \ge 0$ when u is a solution of (9.1). (Here we have subtracted the value of H at its minimum to make the minimum value 0.) To show the generality of this computation and to simplify notation we will write (9.1) as

$$\frac{\partial u_i}{\partial t} = \Delta u_i + f_i(u).$$

Differentiating and using the previous equation gives

$$\frac{\partial h}{\partial t} = \sum_i \frac{\partial H}{\partial u_i} \frac{\partial u_i}{\partial t} = \sum_i \frac{\partial H}{\partial u_i} \cdot (\Delta u_i + f_i(u))$$

$$\frac{\partial^2 h}{\partial x_m^2} = \sum_i \frac{\partial H}{\partial u_i} \frac{\partial^2 u_i}{\partial x_m^2} + \sum_{i,j} \frac{\partial^2 H}{\partial u_i \partial u_j} \frac{\partial u_i}{\partial x_m} \frac{\partial u_j}{\partial x_m}$$

Here and in what follows the indices i and j are summed from 1 to 2. Summing the second equation from $m = 1$ to d gives

$$\Delta h = \sum_i \frac{\partial H}{\partial u_i} \Delta u_i + \sum_{m,i,j} \frac{\partial^2 H}{\partial u_i \partial u_j} \frac{\partial u_i}{\partial x_m} \frac{\partial u_j}{\partial x_m}$$

so using $H_t = \sum_i \frac{\partial H}{\partial u_i} f_i(u)$ gives

$$\frac{\partial h}{\partial t} = \Delta h + H_t - \sum_{m,i,j} \frac{\partial^2 H}{\partial u_i \partial u_j} \frac{\partial u_i}{\partial x_m} \frac{\partial u_j}{\partial x_m}$$

Since H is convex the last term (including the minus sign) is nonpositive and we have

(9.5)
$$\frac{\partial h}{\partial t} \le \Delta h + H_t$$

To prove Theorem 9.4, we will use (9.5) to conclude

(9.6)
$$\sup_{\|x\| \le \epsilon t} h(t,x) \to 0$$

If we were on a bounded set with Neumann boundary conditions this would be easy, since in this case $\inf u_i(t,x) > 0$ at positive times and thus $h(t,x)$ is bounded. If we let x_t be a place where $m_t = \max_x h(t,x)$ is attained then $\Delta h(t,x_t) \le 0$ so

$$\frac{dm_t}{dt} \le H_t \le \sup\{H_t(v) : H(v) = m_t\} < 0$$

and an argument like the one in the proof of Theorem 5.1 shows (9.6).

To prove (9.6) on $\mathbf{R}^d$ we have to deal with the fact that $h(t,x)$ may be unbounded. To do this we first get bounds on how fast the H_t will push h to 0 and then get *a priori* bounds on $h(t,x)$ inside $\|x\| \le at$ that will allow us to drive h to 0. To get upper bounds on H_t (recall it is ≤ 0), we let

$$g(h) = \inf\{-H_t(v_1, v_2) : H(v_1, v_2) \ge h\}$$

We have defined g this way to make it clear that $h \to g(h)$ is increasing. To determine the behavior as $h \to 0$ we observe that at (ρ_1, ρ_2) $\partial H / \partial v_i = 0$ and

$$\frac{\partial^2 H}{\partial v_1^2} = \frac{\beta_2 \rho_1}{v_1^2} \qquad \frac{\partial^2 H}{\partial v_2^2} = \frac{(\beta_1 + \beta_2) \rho_2}{v_2^2} \qquad \frac{\partial^2 H}{\partial v_1 \partial v_2} = 0$$

So near (ρ_1, ρ_2)

$$H(v_1, v_2) - H(\rho_1, \rho_2) \approx \frac{\beta_2}{\rho_1}(v_1 - \rho_1)^2 + \frac{(\beta_1 + \beta_2)}{\rho_2}(v_2 - \rho_2)^2$$

and it follows from (9.4) that $g(h) \sim Bh$. Since $g(h)$ is increasing we have

(9.7) $$g(h) \geq \alpha h/(1+h) \quad \text{for some } \alpha > 0$$

The next step in bounding $h(t, x)$ is to examine the behavior of

$$w' = \frac{-\alpha w}{1 + w} \qquad w(0) = W$$

Since $w(t) \geq W - \alpha t$ the time to reach $\eta > 0$ is at least $(W - \eta)/\alpha$. To see this estimate is fairly sharp observe that while $w(t) \geq W^{1/2} - 1$ we have $w' \leq -\alpha(1 - W^{-1/2})$ so $w(t)$ reaches $W^{1/2} - 1$ at time $\leq W/(\alpha(1 - W^{-1/2}))$. When $w(t) \leq W^{1/2} - 1$ we have $w' \leq -\alpha w/W^{1/2}$ so the time to go from $W^{1/2} - 1$ to η is at most $W^{1/2}\alpha^{-1}(\log(W^{1/2}) - \log \eta)$. Adding the two estimates we see that

(9.8) For $W \geq 4$ and $\eta < 1$ the time to reach η is smaller than

$$2\alpha^{-1}W + CW^{1/2}(\log W - \log \eta).$$

To get *a priori* bounds on h note that our hypotheses imply that $u_i(1, x)$ is positive and continuous so there are constants μ_i so that $u_i(1, x) \geq \mu_i$ for all x with $\|x\|_2 \leq 1$, and we can without loss of generality assume that the last conclusion holds at time 0.

(9.9) **Lemma.** There is a constant K so that if $\|x\|_2 \leq at$ and $t \geq 1$ then $h(t, x) \leq Kt$.

Proof: We have supposed that $\phi_1(x) + \phi_2(x) \leq 1$ so the probabilistic interpretation implies $u_1(t, x) + u_2(t, x) \leq 1$ for all t and x and it follows that

$$\frac{\partial u_1}{\partial t} \geq \Delta u_1 - (\beta_1 + \beta_2)u_1$$

$$\frac{\partial u_2}{\partial t} \geq \Delta u_2 - (\delta_2 + \gamma)u_2$$

To see these inequalities it is convenient to write the right–hand side in the form in (9.2). (Recall that our main assumption (9.3) implies $\beta_1 > \delta_1$.) Let $c_1 = (\beta_1 + \beta_2)$ and $c_2 = \delta_2 + \gamma$. Recalling that solutions of

$$\frac{\partial u}{\partial t} = \Delta v - cv \qquad u(0, x) = \phi(x)$$

are given by

$$u(t, x) = e^{-ct} \int (4\pi t)^{-d/2} e^{-\|x - y\|_2^2/4t} \phi(y) \, dy$$

and using the maximum principle (see (9.11) at the end of this section) we have that when $\|x\|_2 \leq at$

$$u_i(t, x) \geq e^{-c_i t} \mu_i \int_{\|y\| \leq 1} (4\pi t)^{-d/2} e^{-(\|x\|_2 + 1)^2/4t} dy$$

$$\geq C_d \mu_i (4\pi t)^{-d/2} \exp(-(c_i + (a^2/4))t - a/2 - (1/4t))$$

where C_d is the volume of $\{y : \|y\|_2 \leq 1\}$. Combining the last expression with the fact that

$$(9.10) \qquad\qquad h(x) \leq C(1 - \log(\min_i x_i))$$

completes the proof of (9.9) □

Let $a > 0$ be chosen so that $3a^{-1}aK < (1 - a)$, i.e., so that if $\omega(t)$ solves

$$\omega' = -\alpha\omega/(1+\omega) \quad \text{with} \quad \omega(0) = Kat$$

then for any $\eta > 0$ when $t > T_\eta$, $\omega((1-a)t) < \eta$. We will prove Theorem 9.4 with $\sigma = a/2$. Let $D_r = \{y : \|y\|_2 < r\}$ and define $h_1^t(t,x)$ to be the solution of

$$\frac{\partial h}{\partial t} = \Delta h - \alpha h/(1+h) \quad \text{in } \mathcal{D}_t \equiv [at, t] \times D_{at}$$

$$h(s, x) = Ks \quad \text{if } s = at, \text{ or } x \in \partial D_{at}$$

Since $h(s, x) \leq h_1^t(s, x)$ when $s = at$ or $x \in \partial D_{at}$, and $g(h) \geq \alpha h/(1+h)$ it follows from the maximum principle that $h(s, x) \leq h_1^t(s, x)$ for $(s, x) \in \mathcal{D}_t$.

To bound $h_1^t(t, x)$ we will use $h_2^t(s, x) = \omega(s - at)$. Another use of the maximum principle shows $h_1^t(s, x) \geq h_2^t(s, x)$ in $\mathcal{D}_t$. The last inequality is the opposite of the one we want but we will turn it around by showing that the difference is small when $\|x\|_2 \leq at/2$. Intuitively the difference is only due to paths $(t - s, B_s)$ that escape from the space time cylinder $[at, t] \times D_{at}$ on the side. Here B_s is a Brownian motion run at twice its usual speed. When the starting point $\|B_0\|_2 \leq at/2$ this event has exponentially small probability and brings a "reward" $\leq Kt$ so the difference $h_1^t - h_2^t$ goes to 0 exponentially fast as $t \to \infty$.

To begin to turn our intuition into a proof, we let $\bar{g}(x) = \alpha x/(1+x)$ and observe that Itô's formula implies that if $\tau = \inf\{s : B_s \notin D_{at}\}$ then

$$h_i^t(t - (s \wedge \tau), B_{s \wedge \tau}) - \int_0^{s \wedge \tau} \bar{g}(h_i^t(t - r, B_r))\, dr \qquad s \leq (1 - a)t$$

is a bounded martingale. Using the martingale property at time $s = (1 - a)t$ gives

$$h_i^t(t, x) = E_x \left(h_i^t(at, B_{(1-a)t}) - \int_0^{(1-a)t} \bar{g}(h_i^t(t - r, B_r))\, dr \,;\, \tau > (1 - a)t \right)$$

$$+ E_x \left(h_i^t(t - \tau, B_\tau) - \int_0^\tau \bar{g}(h_i^t(t - r, B_r))\, dr \,;\, \tau \leq (1 - a)t \right)$$

Since $h_1^t(at, x) = h_2^t(at, x)$, $0 \leq h_2^t(t-r, B_r) \leq h_1^t(t-r, B_r)$ when $\tau \geq r$, and $\bar{g}$ is increasing, it follows that on $\{\tau > (1 - a)t\}$ we have

$$h_1^t(at, B_{(1-a)t}) - \int_0^{(1-a)t} \bar{g}(h_1^t(t - r, B_r))\, dr$$

$$\leq h_2^t(at, B_{(1-a)t}) - \int_0^{(1-a)t} \bar{g}(h_2^t(t - r, B_r))\, dr$$

Subtracting the two expressions for $h_i^t(t,x)$ and recalling $h_i^t \geq 0$ and $\bar{g} \geq 0$ gives

$$h_1^t(t,x) - h_2^t(t,x) \leq 0 + E_x\left(h_1^t(t-\tau, B_\tau) + \int_0^\tau \bar{g}(h_2^t(t-r, B_r))\,dr; \tau \leq (1-a)t\right)$$

$$\leq (Kt + \alpha t)P_x(\tau \leq (1-a)t)$$

since $h_1^t(s,x) = Ks$ when $x \in \partial D_{at}$ and $0 \leq \bar{g} \leq \alpha$. Standard large deviations estimates for Brownian motion imply that for $\|x\|_2 \leq at/2$, $P_x(\tau < (1-a)t) \leq C\exp(-\delta t)$, so

$$\sup_{\|x\| \leq at/2} |h_1^t(t,x) - h_2^t(t,x)| \to 0 \quad \text{as } t \to \infty$$

Since $h_1^t(t,x) \geq h(t,x)$ for $x \in D_{at}$ and $h_2^t(t,x) = \omega((1-a)t) < \eta$ for $t > T_\eta$, Theorem 9.4 follows. □

For completeness we give a proof of

(9.11) **Maximum Principle.** Suppose $f_1(h) \geq f_2(h)$ and the h_i solve

$$\frac{\partial h_i}{\partial t} = \Delta h_i - f_i(h_i) \quad \text{in } \mathcal{D}_t$$

with $h_1(s,x) \leq h_2(s,x)$ if $s = at$, or $x \in \partial D_{at}$ then $h_1(s,x) \leq h_2(s,x)$ in $\mathcal{D}_t$.

PROOF: This is easier to prove than to find in the library. Suppose first that $f_1(h) > f_2(h)$ and $h_1(s,x) < h_2(s,x)$ if $s = at$, or $x \in \partial D_{at}$. Let s_0 be the smallest value of s for which there is an x with $h_1(s,x) \geq h_2(s,x)$. Continuity of the h_i implies that we can find an x_0 so that $h_1(s_0, x_0) = h_2(s_0, x_0)$. The strict inequality between the h_i on the boundary implies $x_0 \in D_{at}$ and $s_0 > 0$. The definition of s_0 implies that $h_1(s_0, x) \leq h_2(s_0, x)$ for all x. Since $h_1(s_0, x_0) = h_2(s_0, x_0)$, we must have $\nabla h_1(s_0, x_0) = \nabla h_2(s_0, x_0)$ and $\Delta h_1(s_0, x_0) \leq \Delta h_2(s_0, x_0)$. Using the last fact and $f_1(h) > f_2(h)$ it follows that at (s_0, x_0)

$$\frac{\partial h_1}{\partial t} = \Delta h_1 - f_1(h_1) < \Delta h_2 - f_2(h_2) = \frac{\partial h_2}{\partial t}$$

However this implies that $h_1(s_0 - \epsilon, x) > h_2(s_0 - \epsilon, x_0)$ for small ϵ contradicting the definition of s_0, so we must have $h_1(s,x) < h_2(s,x)$ for all $(s,x) \in \mathcal{D}_{at}$. To prove the result in (2.4) now let $f_0(h) = f_1(h) + \epsilon$ and change the boundary values to $h_1(s,x) - \epsilon$. The new solution $h_0^\epsilon(s,x) < h_2(s,x)$ and converges pointwise to $h_1(s,x)$ as $\epsilon \to 0$. □

The main reason for interest in Theorem 9.4 is that it applies to systems of equations. However, as the next two examples suggest we also get interesting information when we apply it to a single equation.

Example 9.2. The basic contact process. If we let $\beta = \lambda N$ where N is the number of neighbors and write u for u_1 then the equation in Example 8.1 can be written as

(9.12) $$\frac{\partial u}{\partial t} = \Delta u - u + \beta(1-u)u \qquad u(0,x) = \phi(x)$$

To find a Lyapunov function we let $\rho = (\beta - 1)/\beta$ and write the dyanmical system as

$$\frac{dv}{dt} = v(-1 + \beta(1 - v)) = \beta v(\rho - v)$$

Taking $H(v) = v - \rho \log v$ and noticing $h'(v) = 1 - (\rho/v)$ we have

$$\frac{dH(v(t))}{dt} = -\beta(v - \rho)^2$$

Clearly H satisfies (9.10). Since $H'(\rho) = 0$ and $H''(v) = \rho/v^2$, repeating the proof of (9.7) shows it is satisfied. Since H is convex we get a convergence result like Theorem 9.4

Theorem 9.5. *Suppose that $\beta > 1$ u solves (9.12) for continuous $0 \le \phi(x) \le 1$ with $\phi(x_0) > 0$ for some x_0. Then there is a $\sigma > 0$ so that as $t \to \infty$,*

$$\sup_{\|x\| \le \sigma t} \left| u(t, x) - \frac{\beta - 1}{\beta} \right| \to 0.$$

Much better convergence results than this are known for this equation (see Aronson and Weinberger (1978) for more general results and Bramson (1983) for more detailed information), but the last result shows that $(\star)$ holds and we have

Theorem 9.6 *Suppose $\beta > 1$. If ϵ is small then the contact process with strring at rate ϵ^{-2} has a translation invariant stationary distribution in which the density of 1's is close to $(\beta - 1)/\beta$.*

Example 9.3. The threshold voter model. In this case if N is the number of neighbors and we write u for u_1 then the limiting equation in Example 8.2 is

$$(9.13) \qquad \frac{\partial u}{\partial t} = \Delta u - u(1 - u^N) + (1 - u)(1 - (1 - u)^N) \qquad u(0, x) = \phi(x)$$

When $N = 1$ the last two terms on the right hand side cancel so we will suppose that $N \ge 2$. For our Lyapunov function we take $H(v) = -\log v - \log(1 - v)$, which has

$$H'(v) = -\frac{1}{v} + \frac{1}{1 - v} = \frac{2v - 1}{v(1 - v)}$$

$$
\begin{aligned}
\frac{dH(v(t))}{dt} &= \frac{2v - 1}{v(1 - v)} \left\{ -v(1 - v^N) + (1 - v)(1 - (1 - v)^N) \right\} \\
&= (2v - 1) \left\{ -(1 + v + \cdots + v^{N-1}) + (1 + (1 - v) + \cdots + (1 - v)^{N-1}) \right\} \\
&= -(2v - 1)^2 \left\{ 1 + \sum_{j=2}^{N-1} \frac{(1 - v)^j - v^j}{1 - 2v} \right\}
\end{aligned}
$$

where the sum is 0 if $N = 2$. Since $(1 - v)^j - v^j$ and $1 - 2v$ are both positive on $v < 1/2$ and negative on $v > 1/2$ their quotient is always positive. To compute the value at $v = 1/2$ we note that L'Hopital's rule implies that

$$\lim_{v \to 1/2} \frac{(1-v)^j - v^j}{1 - 2v} = \lim_{v \to 1/2} \frac{-j(1-v)^{j-1} - jv^{j-1}}{-2v} = j2^{-(j-1)}$$

so the term in braces is bounded away from 0 and ∞.

Since $H'(1/2) = 0$ and $H''(1/2) > 0$ it is easy to see as before that (9.7) holds. The other condition (9.10) does not hold as stated since $H(1) = \infty$. However it is easy to see that under suitable assumptions (9.9) holds and we have

Theorem 9.7. *Suppose that $N \geq 2$ u solves (9.13) for continuous $0 \leq \phi(x) \leq 1$ with $\phi(x_0) > 0$ for some x_0 and $\phi(x_1) < 1$ for some x_1. Then there is a $\sigma > 0$ so that as $t \to \infty$,*

$$\sup_{\|x\| \leq \sigma t} |u(t, x) - 1/2| \to 0.$$

Again better convergence results than this are known for this equation (see Aronson and Weinberger (1978) and Fife and McLeod (1977)) but the last result shows that $(\star)$ holds and we have

Theorem 9.8. *Suppose $N \geq 2$. If ϵ is small then the contact process with strring at rate ϵ^{-2} has a translation invariant stationary distribution in which the density of 1's is equal to 1/2.*

We get "equal to 1/2" rather than just "close to 1/2" by starting from product measure with density 1/2 and using the symmetry of the dynamics under interchange of 0's and 1's. Comparing this with Theorems 5.1 and 5.3, the only surprise is that in the nearest neighbor case there is a stationary distribution with fast stirring. We conjecture that the presence of stirring at any positive rate, there is a nontrivial stationary distribution. In support of this conjecture, Figure 9.2 shows a simulation of the nearest neighbor case with stirring rate $= 3$.

Figure 9.2. Threshold voter model, $d = 1$, $\mathcal{N} = \{-1, 1\}$, with stirring at rate 3

Appendix. Proofs of the Comparison Results

In this section we will prove Theorems 4.1, 4.2, and 4.3. The proofs are not beautiful but by now the reader has hopefully been convinced that they are useful. We begin by recalling the set-up and repeating some definitions that were more fully explained in Section 4. Let

$$\mathcal{L}_0 = \{(x, n) \in \mathbf{Z}^2 : x + n \text{ is even}, n \geq 0\}$$

and make $\mathcal{L}_0$ into a graph by drawing oriented edges from (x, n) to $(x + 1, n + 1)$ and from (x, n) to $(x - 1, n + 1)$. Given random variables $\omega(x, n)$ that indicate whether the sites are open (1) or closed (0), we say that (y, n) can be reached from (x, m) and write $(x, m) \to (y, n)$ if there is a sequence of points $x = x_m, \ldots, x_n = y$ so that $|x_k - x_{k-1}| = 1$ for $m < k \leq n$ and $\omega(x_k, k) = 1$ for $m \leq k \leq n$. We say that the $\omega(x, n)$ are "M dependent with density at least $1 - \gamma$" if whenever (x_i, n_i), $1 \leq i \leq I$ is a sequence with $\|(x_i, m_i) - (x_j, m_j)\|_\infty > M$ if $i \neq j$ then

$$(A.1) \qquad P(\omega(x_i, n_i) = 0 \text{ for } 1 \leq i \leq I) \leq \gamma^I$$

Let $\mathcal{C}_0 = \{(y, n) : (0, 0) \to (y, n)\}$ be the set of all points in space-time that can be reached by a path from $(0, 0)$. $\mathcal{C}_0$ is called the *cluster containing the origin*. When the cluster is infinite, i.e., $\{|\mathcal{C}_0| = \infty\}$ we say that *percolation occurs*. Our first result shows that if the density of open sites is high enough then percolation occurs.

Theorem A.1. If $\theta \leq 6^{-4(2M+1)^2}$ then $P(|\mathcal{C}_0| < \infty) \leq 55\,\theta^{1/(2M+1)^2} \leq 1/20$.

PROOF: The proof is by the contour method. Even though the argument is messy to write down, the idea is simple: if $|\mathcal{C}_0| < \infty$ then there is a "contour" of closed sites that stops the percolation from occuring. As we will show, the probability of a specific contour of length n is $\leq g(\theta)^n$ where $g(\theta) \to 0$ as $\theta \to 0$ and the number of contours of length n is $\leq 3^n$ so by summing a geometric series we see that the existence of a contour is unlikely if θ is small.

Most of the work goes into defining the contour. Before starting on this we have to discard a trivial case: if $(0, 0)$ is closed, an event with probability $\leq \gamma$, then $\mathcal{C}_0 = \emptyset$. For the rest of the proof we will concentrate on the case in which $(0, 0)$ is open and hence $(0, 0) \in \mathcal{C}_0$. Let $D = \{z \in \mathbf{R}^2 : \|z\|_1 \leq 1\}$, where D is for diamond. To turn the cluster $\mathcal{C}_0$ into a solid blob, we look at

$$\mathcal{D}_0 = \cup_{(m,n) \in \mathcal{C}_0} ((m, n) + D)$$

where $(m, n) + D = \{(m, n) + z : z \in D\}$ is the set D translated by (m, n). When $(0, 0) \in \mathcal{C}_0$, the lowest point in $\mathcal{D}_0$ is $(0, -1)$. If $|\mathcal{C}_0| < \infty$, then the open set

$$G = \{\mathbf{R} \times (-1, \infty)\} - \mathcal{D}_0$$

has exactly one unbounded component U. We call $\Gamma = \partial U \cap \mathcal{D}_0$ the *contour* associated with $\mathcal{C}_0$, and orient it so that the segment $(0, -1) \to (1, 0)$, which is always present, is oriented in the direction indicated. For an example see Figure A.1.

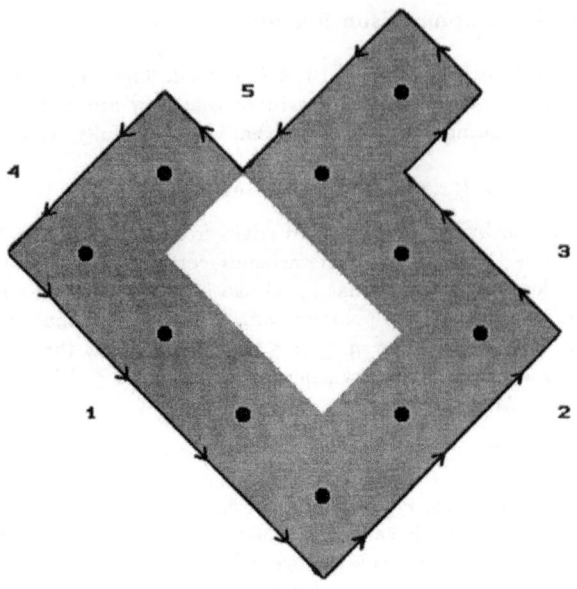

Figure A.1

The contour is made up of segments that are translates of the four sides of D

type	1	2	3	4
translate of	$(-1,0) \to (0,-1)$	$(0,-1) \to (1,0)$	$(1,0) \to (0,1)$	$(0,1) \to (-1,0)$

As we walk along the contour in the direction of the orientation, our left hand is always touching $\mathcal{D}_0$ and our right is always touching U. If we stand at the midpoint of one of the segments that make up Γ then the site in

$$\mathcal{L} = \{(m,n) \in \mathbf{Z}^2 : m + n \text{ is even}\}$$

closest to our right hand is called the *site associated with the segment*. A glance at Figure A.1 reveals that the sites associated with segments of types 3 and 4 must be closed but those associated with types 1 and 2 may be open or closed. Let n_i be the number of segments of type i in the contour. The segments of types 1 and 2 increase the x coordinate by 1, while those of types 3 and 4 decrease the x coordinate by 1. The contour ends where it begins so $n_3 + n_4 = n_1 + n_2$ and hence if the contour is composed of n segments we must have $n_3 + n_4 = n/2$. Now a closed site may be associated with one type 3 and one type 4 segement (see 5 on Figure A.1) but cannot be associated with more than one segment of each type, so if there are n segments in the contour there must be at least $n/4$ closed sites along it.

To count the number of contours of length n we note that the first segment is always $(0, -1) \to (1, 0)$ and after that there are at most 3 choices at each stage (since we cannot retrace the step just made), so there are at most 3^{n-1} contours of length n. Suppose for the moment that the states of the sites are independent and open with probability $1 - \gamma$. Noting that the length of the contour is ≥ 4, it follows that

$$P(0 < |\mathcal{C}_0| < \infty) \leq \sum_{n=4}^{\infty} 3^{n-1} \gamma^{n/4} = \frac{1}{3} \cdot \frac{(3\gamma^{1/4})^4}{1 - 3\gamma^{1/4}} = \frac{27\gamma}{1 - 3\gamma^{1/4}}$$

which is $< 1 - \gamma$ if γ is small enough. (Recall $P(\mathcal{C}_0 = \emptyset) \leq \gamma$.) To extend the last result to the M dependent case, note that we can find a subset of the closed sites along the contour of size at least $n/4(2M + 1)^2$ so that for each $z \neq w$ in this set $\|z - w\|_\infty > M$. (Pick any closed site to start, then throw out the $\leq (2M + 1)^2 - 1$ closed sites in our set that are too close to the first one, pick another site, throw out the closed sites too close to it ...) Using (4.1) and noting our assumption on γ implies $3\gamma^{1/4(2M+1)^2} \leq 1/2$ we have

$$P(0 < |\mathcal{C}_0| < \infty) \leq \sum_{n=4}^{\infty} 3^{n-1} \gamma^{n/4(2M+1)^2}$$

$$= \frac{1}{3} \cdot \frac{(3\gamma^{1/4(2M+1)^2})^4}{1 - 3\gamma^{1/4(2M+1)^2}} \leq 54 \, \gamma^{1/(2M+1)^2}$$

Recalling now that $P(\mathcal{C}_0 = \emptyset) \leq \gamma \leq \gamma^{1/(2M+1)^2}$, we have proved Theorem 4.1. $\square$

From the last proof it follows immediately that if we let $|\Gamma|$ denote the number of segments in the contour and assume $\gamma \leq 6^{-4(2M+1)^2}$ then

$$(A.2) \qquad P(L \leq |\Gamma| < \infty) \leq \sum_{n=L}^{\infty} 3^{n-1} \gamma^{n/4(2M+1)^2} = \frac{1}{3} \cdot \frac{(3\gamma^{1/4(2M+1)^2})^L}{1 - 3\gamma^{1/4(2M+1)^2}} \leq 2^{-L}$$

In order to prove the existence of stationary distributions we need results about M dependent oriented percolation starting from the initial configuration W_0^p in which the events $\{x \in W_0^p\}$, $x \in 2\mathbf{Z}$ are independent and have probability p. Let

$$W_n^p = \{y : (x, 0) \to (y, n) \text{ for some } x \in W_0^p\}$$

Theorem A.2. If $p > 0$ and $\gamma \leq 6^{-4(2M+1)^2}$ then

$$\liminf_{n \to \infty} P(0 \in W_{2n}^p) \geq 1 - 55 \, \theta^{1/(2M+1)^2} \geq 19/20$$

Proof: The first step is to look backwards in time to reduce the new problem to the old one solved in (A.2). This is the discrete time version of the duality considered in Section 3. To have the dual process defined for all time, it is convenient to introduce independent random variables $\omega(x, n)$ for $n < 0$ that have $P(\omega(x, n) = 1) = 1 - \gamma$ and look at the percolation process on $\mathcal{L} = \{(x,n) \in \mathbf{Z}^2 : x + n \text{ is even}\}$. Later in the proof we will want to use the fact that $\gamma > 0$, so you should observe that the desired conclusion is trivial when $\gamma = 0$, i.e., all sites are open.

We say that (x,m) can be reached from (y,n) by a dual path (and write $(y,n) \to_*$ (x,m)) if there is a sequence of points $x = x_m, \ldots, x_n = y$ so that $|x_k - x_{k-1}| = 1$ for $m < k \le n$ and $\omega(x_k, k) = 1$ for $m \le k \le n$. It should be clear from the definition that $(x,m) \to (y,n)$ if and only if $(y,n) \to_* (x,m)$, so

$$\{0 \in W_{2n}^p\} = \{(0, 2n) \to_* (x, 0) \text{ for some } x \in W_0^p\}$$

To estimate the right hand side it is convenient to introduce

$$\hat{W}_m^{2n} = \{x : (0, 2n) \to_* (x, 2n - m)\}$$
$$\hat{C}_{(0,2n)} = \{(x,t) : (0, 2n) \to_* (x, t)\}$$

By conditioning on the value of $\hat{W}_{2n}^{2n}$, it is easy to see that

$(A.3)$
$$P(0 \in W_{2n}^p) = 1 - E\left\{(1-p)^{|\hat{W}_{2n}^{2n}|}\right\}$$

so to complete the proof we want to show that if n is large and $\hat{W}_{2n}^{2n} \ne \emptyset$ then $|\hat{W}_{2n}^{2n}|$ is large with high probability. The process $\hat{W}_m^{2n}$ comes from random variables $\omega(x,n)$ that have property $(A.1)$, and the event on the left hand side of $(A.4)$ below implies that the contour associated with $\hat{C}_{(0,2n)}$ has length at least $4n$, so $(A.2)$ implies

$(A.4)$
$$P(\hat{W}_{2n}^{2n} \ne \emptyset, |\hat{C}_{(0,2n)}| < \infty) \le P(4n \le |\Gamma| < \infty) \le 2^{-4n}$$

Now the sites $(x, -1) \in \mathcal{L}$ are independent of those in $\mathcal{L}_0$ and are closed with probability γ so

$(A.5)$
$$P\left(\hat{W}_{2n+1}^{2n} = \emptyset \,\Big|\, 0 < |\hat{W}_{2n}^{2n}| \le \sqrt{n}\right) \ge \theta^{2\sqrt{n}}$$

Combining $(A.4)$ and $(A.5)$ gives

$(A.6)$
$$P(0 < |\hat{W}_{2n}^{2n}| \le \sqrt{n}) \le \frac{P(\hat{W}_{2n}^{2n} \ne \emptyset, |\hat{C}_{(0,2n)}| < \infty)}{P\left(\hat{W}_{2n+1}^{2n} = \emptyset \,\Big|\, 0 < |\hat{W}_{2n}^{2n}| \le \sqrt{n}\right)} \le 2^{-4n}\gamma^{-2\sqrt{n}}$$

Using $(A.3)$ in the first step; then $(A.6)$ and $P(|\hat{W}_{2n}^{2n}| > 0) \ge P(|\hat{C}_{(0,2n)}| = \infty)$ in the second; and finally, Theorem 4.1 in the third we have

$$P(0 \in W_{2n}^p) \ge \left\{1 - (1-p)^{\sqrt{n}}\right\} P(|\hat{W}_{2n}^{2n}| \ge \sqrt{n})$$
$$\ge \left\{1 - (1-p)^{\sqrt{n}}\right\} \left(P(|\hat{C}_{(0,2n)}| = \infty) - 2^{-4n}\gamma^{-\sqrt{n}}\right)$$
$$\ge \left\{1 - (1-p)^{\sqrt{n}}\right\} \left(1 - 55\,\gamma^{1/(2M+1)^2} - 2^{-4n}\gamma^{-\sqrt{n}}\right)$$

which proves the desired result. □

The arguments for the last two results can be extended easily to give the conclusion quoted in Section 6 as (6.1):

Theorem A.3. If $p > 0$ then

$$\liminf_{n \to \infty} P(\{-2K, \dots, 2K\} \cap W_{2n}^p \neq \emptyset) \geq 1 - \epsilon_K$$

where $\epsilon_K \to 0$ as $K \to \infty$.

PROOF: By the reasoning in the proof of Theorem A.2, we have $\{-2K, \dots, 2K\} \cap W_{2n}^p \neq \emptyset$ if and only if there is a path down from some (x, n) with $|x| \leq 2K$ to $(y, 0)$ for some $y \in W_{2n}^p$. To estimate the probability that this occurs we suppose that all the sites $\{-2K + 1, -2K + 3, \dots, 2K - 1\}$ are open at time $2n + 1$, let

$$\hat{\mathcal{C}} = \{(x, t) : (y, 2n + 1) \to_* (x, t) \text{ for some } |y| \leq 2K - 1\}$$

and turn the cluster $\hat{\mathcal{C}}$ into a solid blob by looking at

$$\hat{\mathcal{D}} = \cup_{(m,n) \in \hat{\mathcal{C}}} (m, n) + D$$

where $D = \{z \in \mathbf{R}^2 : \|z\|_1 \leq 1\}$. As in the proof of Theroem A.1 when $|\hat{\mathcal{C}}| < \infty$ we can define a contour associated with the cluster, and when the contour has length n there will be at least $n/4(2M + 1)^d$ closed sites so that for each $z \neq w$ in this set $\|z - w\|_\infty > M$. Since this time the shortest contour has length $8K$ using $(A.2)$ gives

$$P(|\hat{\mathcal{C}}| < \infty) \leq 2^{-8K}$$

If we let

$$\hat{W}_m^{K,2n+1} = \{y : (x, 2n + 1) \to_* (y, 2n - m) \text{ for some } |x| \leq 2K - 1\}$$

then the argument in the proof of Theorem A.2 shows that

$$P(0 < |\hat{W}_m^{K,2n+1}| \leq \sqrt{n}) \leq 2^{-4n} \gamma^{-\sqrt{n}}$$

So repeating the last computation in the proof of Theorem A.2 proves the result with $\epsilon_K = 2^{-8K}$ □

Our last task is to prove Theorem 4.3. We begin by recalling the

Comparison Assumptions. We suppose given the following ingredients: a translation invariant finite range process $\xi_t : \mathbf{Z}^d \to \{0, 1, \dots \kappa - 1\}$ that is constructed from the graphical representation given in Section 2, an integer L, and a collection H of configurations determined by the values of ξ on $[-L, L]^d$ with the following property:

if $\xi \in H$ then there is an event G_ξ measurable with respect to the graphical representation in $[-k_0 L, k_0 L]^d \times [0, j_0 T]$ and with $P(G_\xi) \geq (1 - \theta)$ so that if $\xi_0 = \xi$ then on G_ξ, ξ_T lies in $\sigma_{2Le_1} H$ and in $\sigma_{-2Le_1} H$.

Here $(\sigma_y \xi)(x) = \xi(x + y)$ denote the translation (or shift) of ξ by y and $\sigma_y H = \{\sigma_y \xi : \xi \in H\}$. If we let $M = \max\{j_0, k_0\}$ then the space time regions

$$\mathcal{R}_{m,n} = (m2Le_1, nT) + \{[-k_0 L, k_0 L]^d \times [0, j_0 T]\}$$

that correspond to points $(m, n), (m', n') \in \mathcal{L}$ with $\|(m, n) - (m', n')\|_\infty > M$ are disjoint.

Theorem A.4. If the comparison assumptions hold then we can define random variables $\omega(x, n)$ so that $X_n = \{m : (m, n) \in \mathcal{L}_0, \xi_{nT} \in \sigma_{m2Le_1} H\}$ dominates an M dependent oriented percolation process with initial configuration $W_0 = X_0$ and density at least $1 - \gamma$, i.e., $X_n \supset W_n$ for all n.

PROOF: We will define the $\omega(x, n)$ in the oriented percolation by induction. We begin by setting $V_0 = X_0$ and defining a slightly enlarged version of the percolation process V_n consisting of all the y so that can be reached from some $(x_0, 0)$ with $x_0 \in V_0$ by a sequence $x_0, x_1, \ldots x_n = y$ so that $|x_k - x_{k-1}| = 1$ for $1 \leq k \leq n$ and $\omega(x_k, k) = 1$ for $0 \leq k < n$, i.e., the last point in the sequence does not have to be open. Since $V_n \supset W_n$ it suffices to show that $X_n \supset V_n$.

Let $n \geq 0$ and suppose that V_n and the $\omega(x, \ell)$ with $\ell < n$ have been defined so that $X_n \supset V_n$. To define the $\omega(m, n)$, and hence V_{n+1}, we consider two cases.

CASE 1. $m \in V_n \subset X_n$. We set $\omega(m, n) = 1$ if $G_{\sigma_{-m2Le_1} \xi_{nT}}$ occurs in the graphical representation translated by $-m2Le_1$ in space and $-nT$ in time, 0 otherwise. By assumption this event is determined by the Poisson points in $\mathcal{R}_{m,n}$, has probability at least $1 - \gamma$, and guarantees that $(m + 1), (m - 1) \in X_{n+1}$.

CASE 2. $m \notin V_n$. In this case, the value of $\omega(m, n)$ is not important for the evolution of the percolation process so we set $\omega(m, n)$ equal to an independent random variable that is 1 with probability $1 - \gamma$ and 0 with probability γ.

If $m \in V_{n+1}$ then either $m - 1 \in V_n$ and $\omega(m - 1, n) = 1$ or $m + 1 \in V_n$ and $\omega(m + 1, n) = 1$. In either case the observation in Case 1 implies that $m \in X_{n+1}$. The last conclusion and induction imply that $X_n \supset V_n$ for all n. The last detail to check is that the $\omega(m, n)$ satisfy $(A.1)$ and again we use induction. If $I = 1$ the conclusion is true, so suppose now that $k > 1$ and that the conclusion is true for $I = k - 1$. Let (x_i, n_i) $1 \leq i \leq k$ be a seqeunce of points with $\|(x_i, n_i) - (x_j, n_j)\|_\infty > M$ if $i \neq j$ and suppose that the sequence has been indexed so that $n_k \geq n_j$ for all $j < k$. Let $\mathcal{F}$ be the information contained in the graphical representation up to time $n_k T$ or in one of the space time boxes $\mathcal{R}_{m_i, n_i}$ with $i < k$. The comparison assumptions and the fact that $n_k \geq n_j$ for $j < k$ imply that

$$P(\omega(m_k, n_k) = 0 | \mathcal{F}) \leq \gamma$$

Integrating the last inequality over $E_{k-1} = \{\omega(m_i, n_i) = 0 \text{ for } i \leq k-1\} \in \mathcal{F}$ which by induction has probability smaller than γ^{k-1} gives

$$\gamma^k \geq \int_{E_{k-1}} \gamma \, dP \geq \int_{E_{k-1}} P(\omega(m_k, n_k) = 0 | \mathcal{F}) \, dP$$

$$= P(E_{k-1} \cap \{\omega(m_k, n_k) = 0\}) = P(E_k)$$

which verifies $(A.1)$ and completes the proof. $\qquad\qquad\square$

REFERENCES

Andjel, E.D. T.M. Liggett, and T. Mountford (1992) Clustering in one dimensional threshold voter models. *Stoch. Processes Appl.* **42**, 73–90

Aronson, D.G. and H.F. Weinberger (1978) Multidimensional diffusion equations arising in population genetics. *Advances in Math.* **30**, 33–76

Asmussen, S. and N. Kaplan (1976) Branching random walks, I. *Stoch. Processes Appl.* **4**, 1–13

Bezuidenhout, C. and L. Gray (1993) Critical attractive spin systems. *Ann. Probab.*, to appear

Bezuidenhout, C. and G. Grimmett (1990) The critical contact process dies out. *Ann. Probab.* **18**, 1462–1482

Bezuidenhout, C. and G. Grimmett (1991) Exponential decay for subcritical contact and percolation processes. *Ann. Probab.* **19**, 984–1009

Boerlijst, M.C. and P. Hogeweg (1991) Spiral wave structure in pre-biotic evolution: hypercycles stable against parasites. *Physica D* **48**, 17–28

Bramson, M. (1983) Convergence of solutions of the Kolmogorov equation to travelling waves. *Memoirs of the AMS*, **285**

Bramson, M. and R. Durrett (1988) A simple proof of the stability theorem of Gray and Griffeath. *Probab. Th. Rel. Fields* **80**, 293–298

Bramson, M., R. Durrett, and G. Swindle (1989) Statistical mechanics of Crabgrass. *Ann. Prob.* **17**, 444–481

Bramson, M. and L. Gray (1992) A useful renormalization argument. Pages ??? in *Random Walks, Brownian Motion, and Interacting Particle Systems*, edited by R. Durrett and H. Kesten, Birkhauser, Boston

Bramson, M. and D. Griffeath (1987) Survival of cyclic particle systems. Pages 21–30 in *Percolation Theory and Ergodic Theory of Infinite Particle Systems* edited by H. Kesten, IMA Vol. 8, Springer

Bramson, M. and D. Griffeath (1989) Flux and fixation in cyclic particle systems *Ann. Probab.* **17**, 26–45

Bramson, M. and C. Neuhauser (1993) Survival of one dimensional cellular automata. Preprint

Chen, H.N. (1992) On the stability of a population growth model with sexual reproduction in Z^2. *Ann. Probab.* **20**, 232–285

Cox, J.T. (1988) Coalescing random walks and voter model consensus times on the torus in Z^d. *Ann. Probab.*

Cox, J.T. and R. Durrett (1988) Limit theorems for the spread of epidemics and forest fires. *Stoch. Processes Appl.* **30**, 171–191

Cox, J.T. and R. Durrett (1992) Nonlinear voter models. Pages 189–202 in *Random Walks, Brownian Motion, and Interacting Particle Systems*, edited by R. Durrett and H. Kesten, Birkhauser, Boston

Cox, J.T. and D. Griffeath (1986) Diffusive clustering in the two dimensional voter model. *Ann. Probab.* **14**, 347–370

DeMasi, A., P. Ferrari, and J. Lebowitz (1986) Reaction diffusion equations for interacting particle systems. *J. Stat. Phys.* **44**, 589–644

DeMasi, A. and E. Presutti (1991) *Mathematical Methods for Hydrodynamic Limits*. Lecture Notes in Math **1501**, Springer, New York

Durrett, R. (1980) On the growth of one dimensional contact processes. *Ann. Probab.* **8**, 890–907

Durrett, R. (1984) Oriented percolation in two dimensions. *Ann. Probab.* **12**, 999–1040

Durrett, R. (1988) *Lecture Notes On Particle Systems And Percolation*. Wadsworth, Belmont, CA

Durrett, R. (1991a) Stochastic models of growth and competition. Pages 1049–1056 in *Proceedings of the International Congress of Mathematicians, Kyoto*, Springer, New York

Durrett, R. (1991b) The contact process, 1974–1989. Pages 1–18 in *Proceedings of the AMS Summer seminar on Random Media*. Lectures in Applied Math **27**, AMS, Providence, RI

Durrett, R. (1991c) Some new games for your computer. *Nonlinear Science Today* Vol. 1, No. 4, 1–7

Durrett, R. (1992a) Multicolor particle systems with large threshold and range. *J. Theoretical Prob.*, 5 (1992), 127–152

Durrett, R. (1992b) A new method for proving the existence of phase transitions. Pages 141-170 in *Spatial Stochastic Processes*, edited by K.S. Alexander and J.C. Watkins, Birkhauser, Boston

Durrett, R. (1992c) Stochastic growth models: bounds on critical values. *J. Appl. Prob.* **29**

Durrett, R. (1992d) *Probability: Theory and Examples*. Wadsworth, Belmont, CA

Durrett, R. (1993) Predator-prey systems. Pages 37–58 in *Asymptotic Problems in Probability Theory: Stochastic Models and Diffusions on Fractals*, edited by K.D. Elworthy and N. Ikeda, Pitman Research Notes in Math 283, Longman, Essex, England

Durrett, R. and D. Griffeath (1993) Asymptotic behavior of excitable cellular automata. Preprint

Durrett, R. and S. Levin (1993) Stochastic spatial models: A user's guide to ecological applications. *Phil. Trans. Roy. Soc. B*, to appear

Durrett, R. and A.M. Moller (1991) Complete convergence theorem for a competition model. *Probab. Th. Rel. Fields* **88**, 121–136

Durrett, R. and C. Neuhauser (1991) Epidemics with recovery in $d = 2$. *Ann. Applied Probab.* **1**, 189–206

Durrett, R. and C. Neuhauser (1993) Particle systems and reaction diffusion equations. *Ann. Probab.*, to appear

Durrett, R. and R. Schinazi (1993) Asymptotic critical value for a competition model. *Ann. Applied Probab.*, to appear

Durrett, R. and J. Steif (1993) Fixation results for threshold voter models. *Ann. Probab.*, to appear

Durrett, R. and G. Swindle (1991) Are there bushes in a forest? *Stoch. Proc. Appl.* **37**, 19–31

Durrett, R. and G. Swindle (1993) Coexistence results for catalysts. Preprint

Eigen, M. and P. Schuster (1979) *The Hypercycle: A Principle of Natural Self-Organization*, Springer, New York

Fife, P.C. and J.B. McLeod (1977) The approach of solutions of nonlinear diffusion equations to travelling front solutions. *Arch. Rat. Mech. Anal.* **65**, 335–361

Fisch, R. (1990a) The one dimensional cyclic cellular automaton: a system with deterministic dynamics which emulates a particle system with stochastic dynamics. *J. Theor. Prob.* **3**, 311–338

Fisch, R. (1990b) Cyclic cellular automata and related processes. *Physica D* **45**, 19–25

Fisch, R. (1992) Clustering in the one dimensional 3-color cyclic cellular automaton. *Ann. Probab.* **20**, 1528–1548

Fisch, R., J. Gravner, and D. Griffeath (1991) Threshold range scaling of excitable cellular automata. *Statistics and Computing* **1**, 23–39

Fisch, R., J. Gravner, and D. Griffeath (1992) Cyclic cellular automata in two dimensions. In *Spatial Stochastic Processes* edited by K. Alexander and J. Watkins, Birkhauser, Boston

Fisch, R., J. Gravner, and D. Griffeath (1993) Metastability in the Greenberg Hastings model. *Ann. Applied. Probab.*, to appear

Grannan, E. and G. Swindle (1991) Rigorous results on mathematical models of catalyst surfaces. *J. Stat. Phys.* **61**, 1085–1103

Gravner, J. and D. Griffeath (1993) Threshold growth dyanmics. *Transactions A.M.S.*, to appear

Gray, L. and D. Griffeath (1982) A stability criterion for attractive nearest neighbor spin systems on **Z**. *Ann. Probab.* **10**, 67–85

Gray, L. (1987) Behavior of processes with statistical mechanical properties. Pages 131–168 in *Percolation Theory and Ergodic Theory of Infinite Particle Systems* edited by H. Kesten, IMA Vol. 8, Springer

Griffeath, D. (1979) *Additive and Cancellative Interacting Particle Systems.* Lecture Notes in Math **724**, Springer

Harris, T.E. (1972) Nearest neighbor Markov interaction processes on multidimensional lattices. *Adv. in Math.* **9**, 66–89

Harris, T.E. (1976) On a class of set valued Markov processes. *Ann. Probab.* **4**, 175–194

Hassell, M.P., H.N. Comins, and R.M. May (1991) Spatial structure and chaos in insect population dynamics. *Nature* **353**, 255–258

Hirsch, M.W. and S. Smale (1974) *Differential Equations, Dynamical Systems, and Linear Algebra*, Academic Press, New York

Holley, R.A. (1972) Markovian interaction processes with finite range interactions. *Ann. Math. Stat.* **43**, 1961–1967

Holley, R.A. (1974) Remarks on the FKG inequalities. *Commun. Math. Phys.* **36**, 227–231

Holley, R.A., and T.M. Liggett (1975) Ergodic theorems for weakly interacting systems and the voter model. *Ann. Probab.* **3**, 643–663

Holley, R.A., and T.M. Liggett (1978) The survival of contact processes. *Ann. Probab.* **6**, 198–206

Kinzel, W. and J. Yeomans (1981) Directed percolation: a finite size renormalization approach. *J. Phys.* A **14**, L163–L168

Liggett, T.M. (1985) *Interacting Particle Systems.* Springer, New York

Liggett, T.M. (1993) Coexistence in threshold voter models. *Ann. Probab.*, to appear

Neuhauser, C. (1992) Ergodic theorems for the multitype contact process. *Probab. Theory Rel. Fields.* **91**, 467–506

Redheffer, R., R. Redlinger, and W. Walter (1988) A theorem of La Salle-Lyapunov type for parabolic systems. *SIAM J. Math. Anal.* **19**, 121–132

Schonmann, R.H. and M.E. Vares (1986) The survival of the large dimensional basic contact process. *Probab. Th. Rel. Fields* **72**, 387–393

Spohn, H. (1991) *Large Scale Dynamics of Interacting Particle Systems*, Springer, New York

Zhang, Yu (1992) A shape theorem for epidemics and forest fires with finite range interactions. Preprint

EXPOSES 1993

ANSEL Jean-Pascal
    Non arbitrage et couverture des actifs

ASPANDIJAROV Sanjor
    Convergence d'une suite de chaînes de Markov dans un quadrant vers une diffusion avec réflexion à la frontière

ATTAL Stéphane
    Algèbre de semimartingales non commutatives. Crochets droit et oblique non commutatifs

BENASSI Albert
    Opérateurs pseudo-différentielles elliptiques, calcul symbolique et processus gaussiens

BLOZNELIS Mindaugas
    On the clt for multiparameter stochastically continuous processes

BRITTON Tom
    On Clustering and Epidemic Models

BROCKHAUS Oliver
    Mesures de Gibbs et frontière de Martin sur l'espace de Wiener

CARMONA Philippe
    Sur les fonctionnelles exponentielles de certains processus de Lévy

CERF Raphaël
    Algorithmes génétiques (GA) et Freidlin-Wentzell

CHASSAING Philippe et ALILI Smail
    Homogeneization or slow diffusion for random walks with random reflecting barriers

CHAUMONT Loïc
    Conditionnement des processus de Lévy et décomposition de trajectoires : application au cas des processus stables

CLOUET Jean-François
    Déformation d'une onde pénétrant dans un milieu aléatoire

COQUIO Agnès
    Calcul de Malliavin et eqxistence et régularité de la densité d'une probabilité invariante d'une diffusion sur une variété Ricmanicnne compacte

DAW Ibrahima
    Sortie d'un domaine et grandes déviations

DUHEILLE Frédérique
    Théorie de Nevanlinna et mouvement brownien

DUMAS Vincent
    Questions de stabilité dans les réseaux à routes fixes

DUNLOP François
    Limite hydrodynamique pour des interfaces

FRANCOIS Olivier
    Modèles Hopfield

FOURATI Sonia
    Deux applications de la théorie générale des processus aux processus de Lévy

GUIONNET Alice
    Dynamique d'un verre de spins

LAROCHE Etienne
    Hypercontractivité des systèmes de spins

MICLO Laurent
    Algorithmes de recuit sous-admissibles

NOBLE John
    Directed Polymer in Random Medium

OVERBECK Ludger
    Semi-Dirichlet forms and Fleming-Viot processes

PESZAT Szymon
    Exponential tail estimates for infinite-dimensional stochastic convolutions

PIAU Didier
    Mouvement brownien et inégalités isopérimétriques

PONTIER Monique
    Nonlinear filtering with a symmetric space valued discontinuous observation

RAINER Catherine
    Sur l'équation de structure $d|X,X|_t = d_t - X_t^+ \, dX_t$

TOPOLSKI Krzysztof
    Limit theorems for the difference of waiting time and queue lenght

YCART Bernard
    Structures aléatoires en combinatoire

LISTE DES AUDITEURS

Mr. PICARD Jean	Université Blaise Pascal, CLERMONT-FD
Mme PONTIER Monique	Mathématiques, Université d'ORLEANS
Mle RAINER Catherine	Laboratoire de Probabilités, Université PARIS VI
Mr. ROBERT Philippe	INRIA, Le Chesnay, ROCQUENCOURT
Mr. RUIZ DE CHAVEZ Juan	Mathématiques, Université de MEXICO (Mexique)
Mle SAADA Ellen	Département de Mathématiques, Université de ROUEN
Mle SAVONA Catherine	Université Blaise Pascal, CLERMONT-FD
Mr. SCHWAB Christian	Institut de Mathématiques, FRIBOURG (Suisse)
Mme TIBI Danielle	U.F.R. de Mathématiques, Université de PARIS VII
Mr. TOPOLSKI Krzysztof	IMathematics, University of WROCLAW (Pologne)
Mr. VIENS Frederi G.	University of California at IRVINE (U.S.A.)
Mr. YCART Bernard	LMC - IMAG, Université de GRENOBLE

LIST OF PREVIOUS VOLUMES OF THE "Ecole d'Eté de Probabilités"

1971 -   J.L. Bretagnolle   (LNM 307)
          "Processus à accroissements indépendants"
          S.D. Chatterji
          "Les martingales et leurs applications analytiques"
          P.A. MEYER
          "Présentation des processus de Markov"

1973 -   P.A. MEYER   (LNM 390)
          "Transformation des processus de Markov"
          P. PRIOURET
          "Processus de diffusion et équations différentielles
          stochastiques"
          F. SPITZER
          "Introduction aux processus de Markov à paramètres
          dans Z_v"

1974 -   X. FERNIQUE   (LNM 480)
          "Régularité des trajectoires des fonctions aléatoires
          gaussiennes"
          J.P. CONZE
          "Systèmes topologiques et métriques en théorie
          ergodique"
          J. GANI
          "Processus stochastiques de population"

1975   A. BADRIKIAN   (LNM 539)
          "Prolégomènes au calcul des probabilités dans
          les Banach"
          J.F.C. KINGMAN
          "Subadditive processes"
          J. KUELBS
          "The law of the iterated logarithm and related strong
          convergence theorems for Banach space valued random
          variables"

1976   J. HOFFMANN-JORGENSEN   (LNM 598)
          "Probability in Banach space"
          T.M. LIGGETT
          "The stochastic evolution of infinite systems of
          interacting particles"
          J. NEVEU
          "Processus ponctuels"

1977   D. DACUNHA-CASTELLE   (LNM 678)
          "Vitesse de convergence pour certains problèmes
          statistiques"
          H. HEYER
          "Semi-groupes de convolution sur un groupe localement
          compact et applications à la théorie des probabilités"
          B. ROYNETTE
          "Marches aléatoires sur les groupes de Lie"

1978	R. AZENCOTT	(LNM 774)
	"Grandes déviations et applications"	
	Y. GUIVARC'H	
	"Quelques propriétés asymptotiques des produits de matrices aléatoires"	
	R.F. GUNDY	
	"Inégalités pour martingales à un et deux indices : l'espace H^p"	
1979	J.P. BICKEL	(LNM 876)
	"Quelques aspects de la statistique robuste"	
	N. EL KAROUI	
	"Les aspects probabilistes du contrôle stochastique"	
	M. YOR	
	"Sur la théorie du filtrage"	
1980	J.M. BISMUT	(LNM 929)
	"Mécanique aléatoire"	
	L. GROSS	
	"Thermodynamics, statistical mechanics and random fields"	
	K. KRICKEBERG	
	"Processus ponctuels en statistique"	
1981	X. FERNIQUE	(LNM 976)
	"Régularité de fonctions aléatoires non gaussiennes"	
	P.W. MILLAR	
	"The minimax principle in asymptotic statistical theory"	
	D.W. STROOCK	
	"Some application of stochastic calculus to partial differential equations"	
	M. WEBER	
	"Analyse infinitésimale de fonctions aléatoires"	
1982	R.M. DUDLEY	(LNM 1097)
	"A course on empirical processes"	
	H. KUNITA	
	"Stochastic differential equations and stochastic flow of diffeomorphisms"	
	F. LEDRAPPIER	
	"Quelques propriétés des exposants caractéristiques"	
1983	D.J. ALDOUS	(LNM 1117)
	"Exchangeability and related topics"	
	I.A. IBRAGIMOV	
	"Théorèmes limites pour les marches aléatoires"	
	J. JACOD	
	"Théorèmes limite pour les processus"	
1984	R. CARMONA	(LNM 1180)
	"Random Schrödinger operators"	
	H. KESTEN	
	"Aspects of first passage percolation"	
	J.B. WALSH	
	"An introduction to stochastic partial differential equations"	

1985-87	S.R.S. VARADHAN "Large deviations" P. DIACONIS "Applications of non-commutative Fourier analysis to probability theorems H. FÖLLMER "Random fields and diffusion processes" G.C. PAPANICOLAOU "Waves in one-dimensional random media" D. ELWORTHY Geometric aspects of diffusions on manifolds" E. NELSON "Stochastic mechanics and random fields"	(LNM 1362)
1986	O.E. BARNDORFF-NIELSEN "Parametric statistical models and likelihood"	(LNS M50)
1988	A. ANCONA "Théorie du potentiel sur les graphes et les variétés" D. GEMAN "Random fields and inverse problems in imaging" N. IKEDA "Probabilistic methods in the study of asymptotics"	(LNM 1427)
1989	D.L. BURKHOLDER "Explorations in martingale theory and its applications" E. PARDOUX "Filtrage non linéaire et équations aux dérivées partielles stochastiques associées" A.S. SZNITMAN "Topics in propagation of chaos"	(LNM 1464)
1990	M.I. FREIDLIN "Semi-linear PDE's and limit theorems for large deviations" J.F. LE GALL "Some properties of planar Brownian motion"	(LNM 1527)
1991	D.A. DAWSON "Measure-valued Markov processes" B. MAISONNEUVE "Processus de Markov : Naissance, Retournement, Régénération" J. SPENCER "Nine Lectures on Random Graphs"	(LNM 1541)
1992	D. BAKRY "L'hypercontractivité et son utilisation en théorie des semigroupes" R.D. GILL "Lectures on Survival Analysis" S.A. MOLCHANOV "Lectures on the Random Media"	(LNM 1581)

1993   P. BIANE   (LNM 1608)
"Calcul stochastique non-commutatif"
R. DURRETT
"Ten Lectures on Particle Systems"

Springer-Verlag
and the Environment

We at Springer-Verlag firmly believe that an international science publisher has a special obligation to the environment, and our corporate policies consistently reflect this conviction.

We also expect our business partners – paper mills, printers, packaging manufacturers, etc. – to commit themselves to using environmentally friendly materials and production processes.

The paper in this book is made from low- or no-chlorine pulp and is acid free, in conformance with international standards for paper permanency.

Lecture Notes in Mathematics

For information about Vols. 1–1425
please contact your bookseller or Springer-Verlag

GPSR Compliance

*The European Union's (EU) General Product Safety Regulation (GPSR)
is a set of rules that requires consumer products to be safe and our
obligations to ensure this.*

*If you have any concerns about our products, you can contact us on
ProductSafety@springernature.com*

In case Publisher is established outside the EU, the EU authorized
representative is:

Springer Nature Customer Service Center GmbH
Europaplatz 3
69115 Heidelberg, Germany

Batch number: 09624486

Printed by Printforce, the Netherlands